Historical Ecology

SCIENCES

Ecosystems and Environment
Field Directors – Françoise Gaill and Dominique Joly

Historical Ecology, Subject Head – Dominique Joly

Historical Ecology

Learning from the Past to Understand the Present and Forecast the Future of Ecosystems

Coordinated by
Guillaume Decocq

ISTE

WILEY

First published 2022 in Great Britain and the United States by ISTE Ltd and John Wiley & Sons, Inc.

ISTE Ltd
27-37 St George's Road
London SW19 4EU
UK

www.iste.co.uk

John Wiley & Sons, Inc.
111 River Street
Hoboken, NJ 07030
USA

www.wiley.com

Library of Congress Control Number: 2022940278

British Library Cataloguing-in-Publication Data
A CIP record for this book is available from the British Library
ISBN 978-1-78945-090-3

ERC code:
PE10 Earth System Science
PE10_3 Climatology and climate change
PE10_4 Terrestrial ecology, land cover change
LS8 Ecology, Evolution and Environmental Biology
LS8_4 Evolutionary ecology

Contents

1

A General Introduction to Historical Ecology

Guillaume DECOCQ
Jules Verne University of Picardie, Amiens, France

Since the second half of the 19th century, several human and social sciences have increasingly incorporated the environment when interpreting human practices and behavior. At the same time, ecological sciences have paid increasing attention to human factors and history when analyzing biodiversity and functioning of forest ecosystems and landscapes. Historical ecology emerged in this context.

How ecological systems have changed over time and how past human activities impacted these dynamics are long-standing research questions in ecology. But the emergence of historical ecology as a discipline aiming at answering these questions is more recent. Though many definitions have been given in the literature, we can simply define historical ecology, after Russel (1997), as the field of ecology which aims to reconstruct the history of ecosystems and landscapes in order to analyze how past events have impacted present days biodiversity and ecosystem functioning. This includes causes and consequences of interactions between humans and the environment (Beller et al. 2017), and as such, this definition is quite close to the one used by anthropologists: an interdisciplinary research program concerned with comprehending temporal and spatial dimensions in the relationships of human societies to local environments and the cumulative global effects of these relationships (Balée 2006). Historical ecology does not exclude any time period a priori, even if most studies focused on historical times (i.e. for which written records are available, even if the date varies among regions around the world).

Historical Ecology,
coordinated by Guillaume DECOCQ.

1.1. The roots of historical ecology

We can trace the roots of historical ecology back to several "undisciplined" scientists, who paved non-conventional roads from their respective discipline, independently from each other. An extensive review can be found in Szabó's epistemological essay (Szabó 2015). Seeking in the past the drivers of current ecological patterns is an old idea. For example, in the 14th century, Dante already described the historical degradation of Mediterranean forests. Later, in the 18th century, German foresters compared old records of forests to recent ones in order to highlight the degradation of forest ecosystems (Stisser 1737), while in England, Barrington published the first paper about indigenous versus introduced tree species in 1769. Georges Perkins Marsh is usually considered as one of the early pioneers of modern historical ecology, with the publication of his seminal book "*The Earth as Modified by Human Action*" in 1874, though the term "historical ecology" explicitly appeared in Tubbs' book on the New Forest (1968).

Several disciplines have independently contributed to the emergence of historical ecology:

– *Forest history*, sometimes labeled historical biogeography of forests. The French historian Alfred Maury may be considered as the founder of this school, with his remarkable thesis on the French forests from the pre-Roman to the modern times (Maury 1850).

– *Historical geography*, especially through the French geographer Vidal de la Blache, who launched the journal *Annales de Géographie* in 1891. This journal published a large number of regional naturalist monographs focusing on landscapes as human-made structures.

– *Ecology* and, more precisely, the field of *paleoecology*, which aims to reconstruct past ecological communities and their dynamics in response to natural and anthropogenic drivers using proxies such as pollens and macro-remains.

– *Rural history*, for which the French historians Fèbvre and Bloch were among the pioneers via the journal *Annales d'Histoire économique et sociale* they launched in 1929. This journal published a large number of papers dealing with impacts of the environment on human activities and landscape planning.

– *Landscape archaeology*, with the seminal work of Hoskins in England, published in 1955. This school was the first to use innovative tools such as aerial photographs to detect and map archaeological artifacts at broader spatial scales, such as former Roman *villae* in croplands. More recently, *environmental archaeology* extended the scope of landscape archaeology to the reconstruction of the relationships between past societies and the environments they lived in, using a wide range of proxies and covering the whole Quaternary era (Wilkinson and Stevens 2003).

– *Environmental history*, which roots in social science and ethnology, and aims to study human interaction with the natural world over time. Interestingly, environmental history emerged in North America with Nash (1967), and mostly makes use of written sources and oral surveys.

In the second half of the 20th century, these various approaches progressively converged and the old partition between disciplines tended to relax: time makes historical ecology more and more a multidiscipline.

1.2. A multidisciplinary approach of socio-ecosystems

Current research projects in historical ecology most often rely on multidisciplinary approaches, i.e. they involve already long-established disciplines to focus on a joint topic or object. Most often, this object is a given ecosystem or landscape, which is considered a socio-ecosystem (Redman et al. 2004). Socio-ecosystems are hybrid systems: they are patterned by the interaction between natural drivers (e.g. geomorphology, soil types, natural vegetation) and anthropogenic drivers (e.g. landscape planning, land-use, resource exploitation). A typical example is former orchards that experienced secondary succession following abandonment (see Chapter 15). Investigating the history of socio-ecosystems thus requires the analysis of two types of archives: ecological archives (e.g. soils, plant remains) and human archives (e.g. written sources, archaeological remains).

Among ecological archives, soil is probably the most relevant one. Soils can keep the memory of past environmental conditions for a very long time, and provide historical ecologists with a number of physical, chemical and biological proxies (see Chapter 7). Depending on their properties, they can preserve plant remains (e.g. pollen, seeds and fruits, charcoals, phytoliths; e.g. Schoonmaker and Foster 1991; Feiss et al. 2017) and animal remains (Fitzpatrick and Keegan 2007), as well as their DNA (Rawlence et al. 2014; Birks and Birks 2016; see also Chapter 18). Even land-uses that took place a very long time ago left an imprint in soil properties, which often still impact the current vegetation. For example, former Roman settlements usually harbor a luxuriant vegetation dominated by eutrophic, neutrophilous species, while on former Roman cultivated fields, the vegetation is more scattered and dominated by acidophilous species (Dupouey et al. 2002; Plue et al. 2008; see also Chapter 14). Many case studies have been published for a range of regions of Western Europe, all with similar results, so that some plant communities are even used as indicators in field archaeological surveys (Decocq 2004) and can guide targeted archaeological excavations. However, soil is a palimpsest and its biological activity typically clouds the vertical layering of deposits, rendering difficult the reconstruction of accurate time sequences, especially when a single proxy is investigated (e.g. soil charcoals; Feiss et al. (2017), but see Chapter 9). It is thus necessary to integrate several proxies with contrasted spatio-temporal resolution to get accurate time reconstructions (see Chapter 8), and to cross-reference the information with other ecological archives, such as botanical indicators (see Chapters 10–12) and also human archives.

Regarding human archives, the availability of written sources is limited and both their quantity and quality decrease as we go further back in the past. For example, maps and cadasters allow us to reconstruct landscape changes over the last few centuries, but with a low reliability for the most ancient ones (see Chapter 4). Aerial photography allows us to cover only the last few decades but still with important insights into fast-changing landscapes (see Chapter 5). Aside comparing the same site at different times, remote sensing techniques can be used to seek archaeological artifacts over extended areas, by evidencing vegetation (e.g. aerial photography) or microtopographic (e.g. LiDAR) anomalies that the landscape has inherited from past human activities (Challis et al. 2008; Beck 2011, see also Chapter 6). Archaeological excavations greatly help in interpreting past human–ecosystem interactions but are more rarely implemented in historical ecology projects, since they can cover only restricted areas (but see Chapter 16).

1.3. Recent trends in historical ecology

Over the last few decades, studies in historical ecology have shifted from purely qualitative and descriptive to more quantitative and mechanistic. Old concepts have been revisited and novel insights have emerged. A typical example is the concept of *ancient woodlands*. A number of descriptive studies, mostly from British plant ecologists (Rackham 1980; Peterken 1981), evidenced differences in plant species composition between ancient woodlands (i.e. woodlands that have continuously existed since the date of the oldest available map, which usually dates back to the end of the 18th century in most countries; Hermy and Verheyen 2007) and recent woodlands (i.e. woodlands that have established on former agricultural lands, within the last two centuries). Today, the concept of *ancient forest species* has largely permeated vegetation science and conservation biology (see Chapters 10 and 12). Now, historical ecology turns to understand the underlying mechanisms. Experimental studies, mostly conducted in Belgium for Europe and at the Harvard forest for North America, counterintuitively revealed that these compositional differences were primarily explained by species' dispersal capacities, rather than by habitat quality. In other words, ancient forest species such as *Anemone nemorosa*, *Oxalis acetosella* or *Hyacinthoides non-scripta* in Europe hardly colonize recent forest patches, not because they cannot establish but because diaspores hardly disperse from source populations to recently created forest habitats, especially in fragmented landscapes. In comparison, shade-tolerant species that are not dispersal-limited (e.g. *Urtica dioica*, *Galium aparine* or *Veronica hederifolia*) can colonize recent and ancient forests equally well. This colonization capacity has even been quantified by an index ranging from minus 100 to 100 (reviewed in Flinn and Vellend 2005; Hermy and Verheyen 2007). This example illustrates an important step in the development of historical ecology as a science, since process-based hypotheses are now experimentally tested.

Historical ecology also leads to hot debates in ecology. A remarkable example is the controversy about what the primeval European forest used to look like. The dominant idea has long been the *closed-canopy forest hypothesis*, mostly based on the forest dynamics observed in nature reserves, once any significant human pressure is removed, as well as on pollen records. This hypothesis has been convincingly challenged by the *savannah hypothesis* introduced by Rackham (1998) and further supported by a number of arguments reviewed in Vera (2000). The main argument of the latter hypothesis is that many species of big herbivores used to graze in forests (e.g. buffalos, aurochs, wild horses), and likely maintain the forested areas quite open. Most of these species went extinct in the early Medieval times, so that natural succession conducted to closed-canopy forests with a typically shade-tolerant understory flora. Beyond the fact that these hypotheses are still controversial (e.g. Birks 2005), both emerged as a result of historical ecology approaches of forests (see also Chapter 13).

Furthermore, historical ecology tackles timely topics in ecology, for example the impact of global climate changes on ecosystems. Progress in ecoinformatic has rendered possible the use of large database compiling old ecological records. For example, at least since the end of the 19th century, vegetation scientists have accumulated millions of vegetation relevés worldwide. When these relevés can be relocalized, it is possible to resurvey the same plots decades later (e.g. Verheyen et al. 2017). The comparison between old and new relevés then makes it possible to quantify vegetation changes and subsequently infer the drivers of these changes, by using trait-based approaches (e.g. Closset-Kopp et al. 2019) or by computing correlation with measured changes in environmental factors (e.g. Perring et al. 2016). Although inferring processes from resurvey studies is not trivial (see Chapter 2), legacy studies led to important results, for example: the overriding influence of canopy closure on atmospheric nitrogen deposits in explaining vegetation eutrophication (Verheyen et al. 2012); the importance of land-use legacies in explaining the response of forest biodiversity to global change (Perring et al. 2016); and the crucial role of forest management in preserving forest microclimate to buffer against global warming and conserve the forest understory biodiversity (De Frenne et al. 2014; Bernhardt-Römermann et al. 2015; see also Chapter 20). Long-term monitoring studies provide further support to the inferences made from these retrospective analyses (see Chapter 3). Historical ecology has evidenced the inertia of ecosystem in their response to environmental changes and led to novel concepts that largely permeate ecology, for example the concepts of *extinction debt* and *colonization credit* (Kuussaari et al. 2009; Jackson and Sax 2010; see also Chapter 21).

1.4. The way forward

If the roots of historical ecology are in Europe, its ramifications now reach all continents. Several international working groups dedicated to historical ecology emerged, within existing ecological societies such as the International Association for Landscape Ecology (IALE; https://www.landscape-ecology.org/page-18083) and the International Association for Vegetation Science (IAVS; https://www.iavs.org/page/working-groups_historical-vegetation-ecology), and as standalone associations, such as the American Society for Environmental History (https://aseh.org) and its European counterpart (http://eseh.org). This international spread of historical ecology is associated with a diversification of the studied systems. Temperate and Mediterranean forests are still an important focus (see Chapters 3, 9, 10, 11, 12, 14, 15 and 20), but this book shows that historical ecology now concerns all types of ecosystem and landscape: boreo-nemoral forests (Chapter 16), arctic (Chapter 17) and tropical (Chapter 5) vegetation, temperate island (Chapter 18), subtropical coastal dunes (Chapter 19), always with important insights for the understanding of how past interactions between natural ecosystems and human societies have driven current biodiversity patterns and ecosystem functioning.

The challenge for historical ecology since the beginning has been to understand present-day ecosystems by putting them in their historical context. Moreover, historical ecology becomes an applied discipline, for example, to identify reference plant communities needed by restoration ecology practitioners (Swetnam et al. 1999; Egan and Howell 2001; Foster 2002; Jackson and Hobbs 2009) and, more generally, to use the past to predict the future (see Chapter 22). It is thus becoming a field which informs stakeholders and managers. Landscapes and ecosystems cannot be reduced to natural ecological systems disturbed by human activities. Instead, they are dynamic socio-ecosystems in which human activities are both causes and consequences of the observed patterns through time and across spatial scales. This is a timely task at the Anthropocene epoch, since understanding drivers of past ecosystem changes may allow us to predict future changes in response to interacting environmental and human drivers. Time has come to understand the mechanisms behind these changes, in order to open the way for a functional historical ecology. Historical ecology has the potential to become a "transdiscipline" dedicated to the study of complex socio-ecosystem dynamics.

1.5. References

Balée, W. (2006). The research program of historical ecology. *Annu. Rev. Anthropol.*, 35, 75–98.

Beck, A. (2011). Archaeological applications of multi/hyper-spectral data – Challenges and potential. In *Remote Sensing for Archaeological Heritage Management*, Cowley, D. (ed.). Archaeolingua, Budapest.

Beller, E., McClenachan, L., Trant, A., Sanderson, E.W., Rhemtulla, J., Guerrini, A., Grossinger, R., Higgs, E. (2017). Toward principles of historical ecology. *Am. J. Bot.*, 104, 645–648.

Bernhardt-Römermann, M., Baeten, L., Craven, D., De Frenne, P., Hédl, R., Lenoir, J., Bert, D., Brunet, J., Chudomelová, M., Decocq, G. et al. (2015). Drivers of temporal changes in temperate forest plant diversity vary across spatial scales. *Glob. Change Biol.*, 21, 3726–3737.

Birks, H.J.B. (2005). Mind the gap: How open were European primeval forests? *Tr. Ecol. Evol.*, 20, 154–156.

Birks, H.J.B. and Birks, H.H. (2016). How have studies of ancient DNA from sediments contributed to the reconstruction of Quaternary floras? *New Phytol.*, 209, 499–506.

Challis, K., Kokalj, Z., Kincey, M., Moscrop, D., Howard, A.J. (2008). Airborne lidar and historic environment records. *Antiquity*, 82, 1055–1064.

Closset-Kopp, D., Hattab, T., Decocq, G. (2019). Do drivers of forestry vehicles also drive herb layer changes (1970–2015) in a temperate forest with contrasting habitat and management conditions? *J. Ecol.*, 107, 1439–1456.

De Frenne, P., Rodriguez-Sanchez, F., Coomes, D.A., Baeten, L., Verstraeten, G., Vellend, M., Bernhardt-Römermann, M., Brown, C.D., Brunet, J., Cornelis, J. et al. (2014). Forest canopy closure buffers plant community responses to global warming. *Proc. Nat. Acad. Sci. USA*, 110, 18561–18565.

Decocq, G. (2004). Utilisation de la flore et de la végétation actuelles en prospection archéologique. In *Méthodes et initiations d'histoire et d'archéologie*, Racinet, P. and Schwerdroffer, J. (eds). Editions du Temps, Nantes.

Dupouey, J.L., Dambrine, E., Laffite, J.D., Moares, C. (2002). Irreversible impact of past land use on forest soils and biodiversity. *Ecology*, 83, 2978–2984.

Egan, D. and Howell, E.A. (eds) (2001). *The Historical Ecology Handbook: A Restorationist's Guide to Reference Ecosystems.* Island Press, Washington.

Feiss, T., Horen, H., Brasseur, B., Buridant, J., Gallet-Moron, E., Decocq, G. (2017). Historical ecology of lowland forests: Does pedoanthracology support historical and archaeological data? *Quat. Inter.*, 457, 99–112.

Fitzpatrick, S.M. and Keegan, W.F. (2007). Human impacts and adaptations in the Caribbean Islands: An historical ecology approach. *Earth Env. Sci. Trans. Roy. Soc. Edinburgh*, 98, 29–45.

Flinn, K.M. and Vellend, M. (2005). Recovery of forest plant communities in post-agricultural landscapes. *Front. Ecol. Env.*, 3, 243–250.

Foster, D.R. (2002). Conservation issues and approaches for dynamic cultural landscapes. *J. Biogeograph.*, 29, 1533–1535.

Hermy, M. and Verheyen, K. (2007). Legacies of the past in the present-day forest biodiversity: A review of past land-use effects on forest plant species composition and diversity. *Ecol. Res.*, 22, 361–371.

Jackson, S.T. and Hobbs, R.J. (2009). Ecological restoration in the light of ecological history. *Science*, 325, 567–569.

Jackson, S.T. and Sax, D.F. (2010). Balancing biodiversity in a changing environment: Extinction debt, immigration credit and species turnover. *Tr. Ecol. Evol.*, 25, 153–160.

Kuussaari, M., Bommarco, R., Heikkinen R.K., Helm, A., Krauss, J., Lindborg, R., Öckinger, E., Pärtel, M., Pino, J., Rodà F. et al. (2009). Extinction debt: A challenge for biodiversity conservation. *Tr. Ecol. Evol.*, 24, 564–571.

Maury, A. (1850). *Histoire des grandes forêts de la Gaule et de l'ancienne France.* A. Leleux, Paris.

Nash, R.F. (1967). *Wilderness and the American Mind.* Yale University Press, New Haven.

Perring, M.P., De Frenne, P., Baeten, L., Maes, S.L., Depauw, L., Blondeel, H., Caron, M.M., Verheyen, K. (2016). Global environmental change effects on ecosystems: The importance of land-use legacies. *Glob. Change Biol.*, 22, 1361–1371.

Peterken, G.F. (1981). *Woodland Conservation and Management.* Chapman & Hall, London.

Plue, J., Hermy, M., Verheyen, K., Thuillier, P., Saguez, R., Decocq, G. (2008). Persistent changes in forest vegetation and seed bank 1,600 years after human occupation. *Landsc. Ecol.*, 23, 673–688.

Rackham, O. (1980). *Ancient Woodland: Its History, Vegetation and Uses in England.* Edward Arnold, London.

Rackham, O. (1998). Implications of historical ecology for conservation. In *Conservation Science and Action*, Sutherland, W.J. (ed.). Blackwell Science Oxford, Oxford.

Rawlence, N.J., Lowe, D.J., Wood, J.R., Young, J.M., Churchman, G.J., Huang, Y.T., Cooper, A. (2014). Using palaeoenvironmental DNA to reconstruct past environments: Progress and prospects. *J. Quatern. Sci.*, 29, 610–626.

Redman, C., Grove, M.J., Kuby, L. (2004). Integrating social science into the long term ecological research (LTER) network: Social dimensions of ecological change and ecological dimensions of social change. *Ecosystems*, 7, 161–171.

Russell, E.W.B. (1997). *People and the Land through Time: Linking Ecology and History*. Yale University Press, New Haven.

Schoonmaker, P.K. and Foster, D.R. (1991). Some implications of paleoecology for contemporary ecology. *Bot. Rev.*, 57, 204–245.

Stisser, F.U. (1737). *Forst- und Jagd-Historie der Teutschen*. Ritter, Jena.

Swetnam, T.W., Allen, C.D., Betancourt, J.L. (1999). Applied historical ecology: Using the past to manage for the future. *Ecol. Applic.*, 9, 1189–1206.

Szabó, P. (2015). Historical ecology: Past, present and future. *Biol. Rev.*, 90, 997–1014.

Vera, F.W.M. (2000). *Grazing Ecology and Forest History*. CABI Publishing, Wallingford.

Verheyen, K., Baeten, L., De Frenne, P., Bernhardt-Römermann, M., Brunet, J., Cornelis, J., Decocq, G., Dierschke, H., Eriksson, O., Hedl, R. et al. (2012). Driving factors behind the eutrophication signal in understorey plant communities of deciduous temperate forests. *J. Ecol.*, 100, 352–365.

Verheyen, K., De Frenne, P., Baeten, L., Waller, D., Hédl, R., Perring, M., Brunet, J., Chudomelova, M., De Lombaerde, E., Depauw, L. et al. (2017). Combining community resurvey data to advance global change research. *BioScience*, 67, 73–83.

Wilkinson, K. and Stevens, C. (2003). *Environmental Archaeology. Approaches, Techniques and Applications*. Tempus Publishing, Stroud.

2

Historical Resurveys Reveal Causes of Long-term Ecological Change

Donald M. WALLER
University of Wisconsin, Madison, USA

2.1. Serious ecological changes are pervasive

Historical approaches have always informed ecology but have gained great importance now as we struggle to understand the forces driving the many ecological changes occurring around us. These changes are pervasive and accelerating, yet they remain largely hidden as they are difficult to track. Documenting even simple changes in diversity is impossible except where we have historical data. One of Europe's revered natural areas, the Bialowieza forest in Poland, lost 45% of its 133 vascular plant species between 1969 and 1992 (Kwiatkowska 1994). Two protected virgin forests in Pennsylvania lost 59% and 80% of their plant species between 1929 and 1995 (Rooney and Dress 1997). Three state parks in northern Wisconsin, USA, lost 50% of their plant diversity over the past 50 years (Rooney et al. 2004). Middlesex Fells, a 400 ha park near Boston, USA, lost 37% of its 422 plant species between 1894 and 1993 (Drayton and Primack 1996). Although losing species from small particular sites may seem inconsequential, these studies suggest we are losing species from protected temperate forests around the world.

What causes these losses? Are they particular to these sites or occurring in most temperate forests? Are similar forces at work in these cases or is each different? Although these results are dramatic, they would entirely escape notice except for our

Historical Ecology,
coordinated by Guillaume DECOCQ.

having historical data at these sites. Knowing how these forces affect community dynamics improves our ability to identify the particular forces that most threaten biodiversity and is thus essential for wisely managing and restoring these ecosystems.

2.2. Anthropogenic drivers of ecological change

Plant communities respond to many anthropogenic forces. Humans have driven ecological changes for millennia via fire, herding ungulates and tilling fields. However, the scope and rates of anthropogenic environmental change escalated greatly through the 20th century. These forces continue to expand, ensuring that they will increasingly affect ecosystems through the 21st century. Here, I focus on six anthropogenic forces known to greatly affect plant communities. I cite examples from long-term resurveys of sites in Wisconsin, USA, and forestREplot (forestreplot.ugent.be/) resurveys in Europe to illustrate how these forces affect temperate forests. The Wisconsin work capitalizes on high-quality baseline surveys by J.T. Curtis (1959) and his students matched to subsequent resurveys of the same sites 50+ years later (Waller et al. 2012). In temperate forests, most plant species occur in the herb layer (Gilliam et al. 2016). Their shorter stature and life cycles mean that these species respond faster to anthropogenic forces than forest trees.

I first discuss difficulties in studying long-term ecological change. Understanding these help us to explain why good long-term studies are scarce. I then review the different kinds of ecological change that exist and methods to characterize these. These include natural and manipulative experiments, metacommunity models and various statistical approaches. With this background, I then review evidence linking ecological changes to shifts in disturbance regimes, habitat fragmentation, climate change, nitrogen deposition and grazing by herbivores. These concrete examples help us understand the strengths and weaknesses of each approach. Finally, I summarize our current understanding of the forces driving ecological change and hint at new approaches that might help us advance further.

2.2.1. *The missing baseline problem*

Without adequate baseline data, we cannot detect long-term ecological change or analyze how systems have changed (Magnuson 1990). Missing baselines also prevent us from measuring species losses and biotic homogenization, shifts in community composition and other long-term changes. Worse, the absence of data is often taken to mean an absence of effects. In the few cases where we do have detailed baseline data, we often find serious biotic declines (see opening). It behooves us to seek out historical data whenever possible, to archive them carefully and encourage their wider use (Verheyen et al. 2017). Baseline data are often limited to simple species lists as in the opening examples. Such lists identify the species present but cannot inform us about shifts in relative abundance or community structure. It is more useful to have quantitative data on species abundances at several times (Figure 2.1). More extensive data enable more general and robust conclusions about the *kinds* of changes occurring and how *consistent* these changes are. Having data from multiple sites and species also presents challenges; however, in that effects of local circumstances can be complex. Larger datasets provide more power for increasingly sophisticated analyses, allowing us to better understand the patterns and processes of ecological changes.

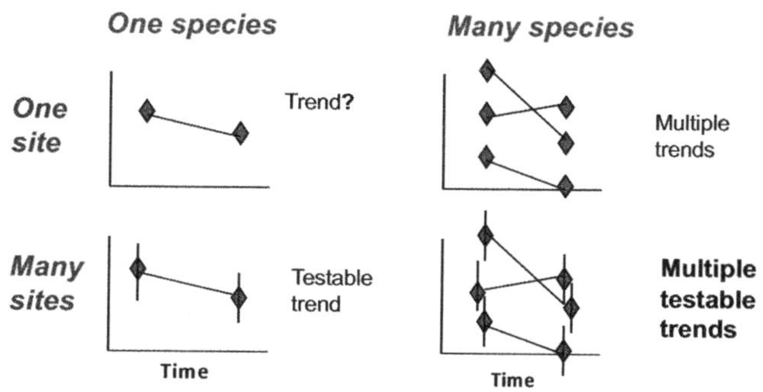

Figure 2.1. *Our ability to characterize ecological change and reliably infer causes rests on having adequate baseline data from many species and sites and a suitable interval between surveys. Such data are rare. For a color version of this figure, see www.iste.co.uk/decocq/ecology.zip*

2.2.2. *Ecological communities are complex*

Many forces drive long-term ecological change, often in combination, complicating our efforts to infer them. These mechanisms operate at different times or places, changing at various scales and rates (Bernhardt-Römermann et al. 2015). Drivers also interact in complex ways, complicating our efforts to infer their relative strengths and patterns of causality. Finally, it is difficult to distinguish systematic patterns of change attributable to any given driver from stochastic changes driven by shifts in local conditions and chance events. Having rich data from many sites (Figure 2.1) gives us the statistical power we need to disentangle multiple drivers while averaging over variable local conditions.

2.3. Kinds of ecological change

Ecologists now use historical data to track many kinds of ecological change. These include changes in species abundance and relative abundance; local colonizations and extirpations; invasions by non-native species; changes in community and habitat structure (e.g. plant density, cover and stature; changes in local (α), among-site (β) and regional (γ) diversity; altered relationships with other species (e.g. the incidence of herbivory and diseases); and shifts in higher-order meta-community properties. At the community level, we track shifts in relative abundance, plant functional traits, species–area relationships, etc. Studies usually report just one or a few types of change, making it difficult to infer how these various kinds of change relate to each other.

2.3.1. *Natural community dynamics*

To detect and analyze the effects of anthropogenic drivers of ecological change, we must be able to separate those from natural processes affecting community dynamics (as with stochastic noise). Classic ecological work has documented how plant communities respond to variation in climate, soils and patterns of disturbance. (e.g. Curtis 1959). Well-described baseline conditions help here, too. European ecologists use Ellenberg indicator values from indicator species to infer environmental conditions (Ellenberg 1991; Bartelheimer and Poschlod 2016). Recurring natural disturbances including tree falls, fires and floods restart succession, generating well-characterized changes in forest plant communities.

2.3.2. *Anthropogenic drivers of ecological change*

2.3.2.1. *Shifts in disturbance regimes*

Farm fields, managed forests, houses, roads and urban areas have displaced prairies, savannas, wetlands and wild forests across much of temperate North America, especially in areas favorable for agriculture. These shifts in land-use greatly reduce natural disturbances like floods and fires, decimating populations of species dependent on these disturbances. In contrast, species adapted to human-disturbed environments thrive. This group includes many invasive species which can themselves displace native species (see section 3.2.5). Species adapted to historically dominant patterns of natural disturbance do poorly follow changes in the type, frequency, scale or intensity of historical disturbance regimes. This undermines the ability of many species to thrive, leading to the pronounced declines in diversity observed when burning declines in fire-dominated systems like prairies (Alstad et al. 2016), savannas and jack-pine-dominated forests on sandy soils (Li and Waller 2015). Floodplain communities are similarly disrupted when flooding either declines or becomes more extreme (Johnson et al. 2016). Recurrent logging similarly threatens many native species by failing to generate the same kinds and amounts of gaps, tip-up mounds, soil structure and coarse woody debris as natural death and tree-falls do in mature temperate forests (Alverson et al. 1994).

2.3.2.2. *Habitat loss and fragmentation*

GIS analyses of aerial imagery allow us to evaluate how landscape features affect patterns of diversity. Forest patches in southern Wisconsin, like those in other temperate areas, are now smaller and more isolated than they once were. This reduces the abundance of species sensitive to edge, area or isolation effects leading to many extirpations. Edge effects penetrate far into mature forests. The densities of roads and houses within a 5 km radius cause native plant diversity to decline and invasions of exotic species to increase (Rogers et al. 2008). As species that need larger areas or less disturbed conditions disappear locally, the diversity of forest patches declines. Over the latter half of the 20th century, this considerably reduced the diversity of smaller forest patches in southern Wisconsin (Figure 2.2). This strengthened the species–area relation matching how land-bridge islands lose diversity after they become isolated, a tenant of island biogeography (MacArthur and Wilson 1967). Smaller forest patches seem destined to continue losing species if they remain small, isolated and subject to other stresses.

2.3.2.3. *Atmospheric nitrogen deposition*

Temperate ecosystems in proximity to industrialized urban areas or intensive agriculture are often affected by atmospheric nitrogen (N) deposition. The wet and dry deposition of oxidized forms of N and sulfur cause “acid rain”, leaching needed cations and mobilizing toxic metals. This harms many species. Farmers apply reduced forms of N (as ammonium compounds) as fertilizer, some of which escapes as drift. Higher ambient levels of biologically available N favor plant species that thrive under high N conditions and disfavor plants (like many Fabaceae) that either fix N or perform well under low nutrient conditions (Verheyen et al. 2012). Such “terrestrial eutrophication” commonly acts to simplify plant communities by favoring fast-growing N-loving plants that can then overtop and outcompete more nutrient-thrifty and shorter-statured plant species (Stevens et al. 2018). Across the temperate U.S., higher N deposition acts to depress plant diversity particularly on low nutrient soils (Simkin et al. 2016). Over half the 348 species evaluated are vulnerable to N deposition including many rare habitat specialists and species with low leaf N (Clark et al. 2019). These results confirm that N deposition has driven widespread changes in many plant communities (Figure 2.3).

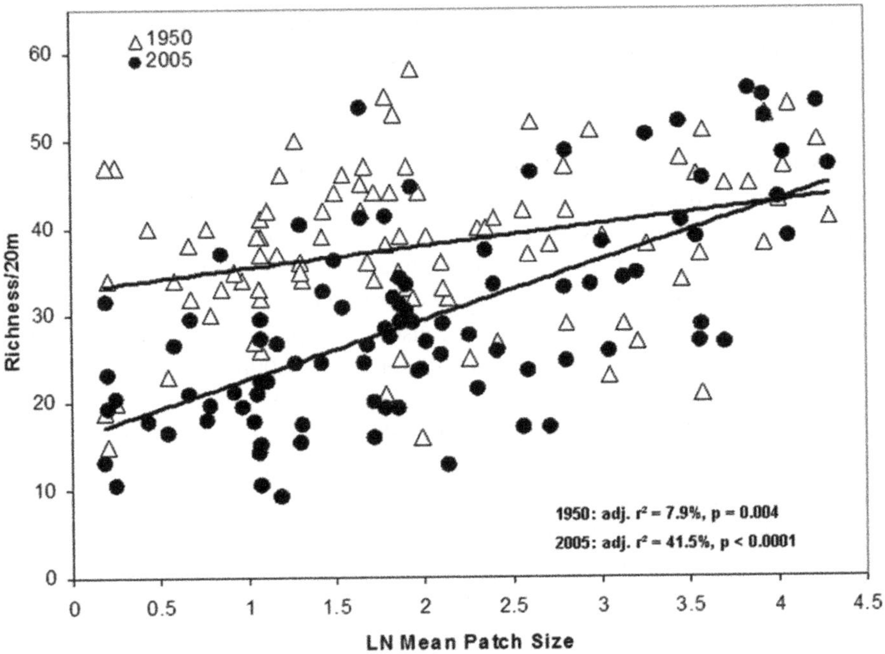

Figure 2.2. *The number of species sustained in smaller forest fragments in southern Wisconsin declined greatly through the latter 20th century. This strengthened the species–area relationship. Data from Rogers et al. (2008)*

2.3.2.4. *Climate change*

Rates of climate change are accelerating, resulting in sizable shifts in rainfall, temperature and seasonality. Shifts in climate are now altering leaf and flowering phenology, shifting elevational and latitudinal ranges, and increasing risks of local extirpation (Parmesan and Yohe 2003). In Wisconsin, night and winter temperatures have increased the most, with local increases in precipitation and more frequent extreme storms (WICCI 2010). These shifts in climate have already shifted the distributions of many species relative to the mid-20th century (Ash et al. 2016).

These rates of movement, however, are slower than those of climate change. This creates a growing mismatch between species distributions and ideal climatic conditions. This lag will likely to increase, threatening the ability of many species to persist.

2.3.2.5. *Invasions by non-native species*

Many species have spread far beyond their original ranges to new regions either by hitching rides with human commerce or via intentional introductions. Some species continue to spread and have become invasive. The absence of coevolved predators and pathogens in their introduced range may play a role in these expansions, particularly when species reallocate resources from defense to competitive ability and enhanced reproduction (Blossey and Notzold 1995). Introduced plants compete for resources and alter the environment, shifting species composition and potentially threatening native species (Callaway and Aschehoug 2000). Prominent invaders of southern Wisconsin forests include *Alliaria petiolata, Berberis thunbergii* and *Rhamnus cathartica.* These species reduce plant diversity, tree regeneration and the nesting success of birds. Introduced insects like the Hemlock Woolly Adelgid and the Emerald Ash Borer are eliminating *Tsuga canadensis* and several *Fraxinus* species from broad regions. Introduced diseases like Chestnut Blight, Dutch Elm Disease and Beech Bark Disease have already massively altered North American forests.

2.3.2.6. *Hyper-abundant herbivores*

Deer (cervids) became far more abundant across much of North America and parts of Europe through the 20th century, with diverse and often dramatic cascading impacts on many understory herbaceous and woody plant species as well as forest structure, soil dynamics, and bird and mammal abundances (Alverson et al. 1988; Waller and Alverson 1997; Côté et al. 2004). In northern Wisconsin, species known to resist or tolerate herbivory by white-tailed deer (grasses, sedges and some ferns) increased dramatically over the past 50 years. In contrast, most native forbs and species known to be susceptible to herbivory (e.g. lilies and orchids) suffered large declines (Balgooyen and Waller 1995). Geographic patterns confirm that deer drive long-term shifts in community composition. Islands lacking deer and Indian reservations with few or no deer still support high diversity including species highly palatable to deer now rare elsewhere. Fenced exclosures confirm how deer radically change the cover, diversity and composition of understory plants including the ability of several late-successional species to effectively recruit saplings. Once deer deplete the abundance of palatable species, the scarcity of seed sources and the tendency of deer to seek out favored species means these species are slow to recover.

2.3.2.7. *Drivers interact*

Although each driver has substantial effects by itself, these drivers also interact. For example, abundant deer tend to favor invasions by weedy plants like *Alliaria* and *Rhamnus* by compacting soils, accelerating nutrient cycling, dispersing seeds and preferentially consuming their native competitors (Dobson and Blossey 2015; Shen et al. 2016). One aspect of widespread biotic homogenization in Europe is the replacement of smaller-ranged plant species by those with larger ranges (Staude et al. 2020). Interestingly, these larger-ranged species have higher leaf N and thus probably benefit from N deposition. Additional key interactions surely exist among the drivers reviewed here. Ignoring such interactions leads us to underestimate the scale and complexity of the forces driving ecological change.

2.4. Understanding the forces driving ecological change

Scientists reduce their uncertainty by formulating hypotheses, conducting experiments and analyzing the results from those experiments to test hypotheses. Given the number and scale of human-related factors affecting ecological systems in the 21st century, the scarcity of baseline data, time delays in these effects and how complex these responses may be, we cannot conduct experiments to manipulate all relevant factors. Instead, ecologists infer these forces using various direct and indirect methods.

2.4.1. *Natural experiments*

Particular patterns of change often give us clues about the mechanisms driving change. We compare sites that differ in local conditions or landscape context to see how these affect the ecological changes observed. In cases where we have systematic differences among regions or locales in environmental conditions, we analyze these differences as "treatments" in a "natural experiment", for example, by comparing watersheds with or without dams (Johnson et al. 2016) or areas with and without invasive earthworms. Comparing islands in Lake Superior with or without deer confirmed their strong influence (Balgooyen and Waller 1995). Indian reservations also provide natural experiments for assessing the effects of alternative paths of forest and wildlife management (Waller and Reo 2018). When many sites are analyzed together that cover a broad range of conditions, we may gain enough power to perform factorial natural experiments to test multiple drivers and their interactions. We can also relate species' functional traits to their within and among-site dynamics (Waller et al. 2017).

2.4.2. *Metrics of change*

By tracking multiple species or whole communities, we can tally local species colonizations and extinctions or increases and decreases in plant size, density or abundance. On the community level, we can track changes in structure and stature as well as species (and trait) composition or overall diversity. Diversity has many components and can be measured in many ways. Spatially, we distinguish among local (α), among-site (β) and overall regional (γ) diversity. Beta diversity reflects differences among biotic communities and thus provides an inverse measure of "biotic homogenization" (McKinney and Lockwood 1999). These concepts and measures have now been sorted into a coherent theory based on Hill numbers (Jost 2007), with natural extensions to measure trait and phylogenetic diversity (Chao et al. 2019).

2.4.3. *Can functional traits reveal drivers of change?*

Plant functional traits have been hailed as a vehicle to "rebuild" community ecology (McGill et al. 2006). Because plant functional traits reflect how plants respond to environmental conditions (Rolhauser et al. 2021), we gain insights into the forces driving shifts in community composition by examining how traits shift as species rise or fall in abundance. Knowing how traits affect physiological and ecological performance under various conditions allows us to predict how changed conditions likely affect species with different trait profiles (Figure 2.3). For example, systematic increases in nitrogen (N) deposition will tend to favor faster-growing plants with higher leaf N and specific leaf area (fitting many invasive plant species). Habitat fragmentation, in contrast, tends to favor species that disperse well, for example, with small seeds or those that attach, or are eaten by, animals. Deer feed selectively on taller plants with thin nutritious leaves, leaving in their wake plants with better physical and chemical defenses. Tracking shifts in community trait profiles allows us to test these predictions. Researchers now use functional traits to track how communities respond to N-enrichment (Suding et al. 2005), climate change (Heilmeier 2019; Thomas et al. 2020), habitat fragmentation (Vellend et al. 2006) and various kinds of disturbance (Mayfield et al. 2010).

DRIVER	Leaf N	SLA	Height	Dispersibility	Tannin Defenses	Leaf Toughness
N-deposition	↑	↑	(↑)			↓
Habitat fragmentation				↑		
Deer herbivory	(↓)	↓	↓		↑	↑

Figure 2.3. *Factors that alter plant communities affect species differently depending on their functional traits. This can generate diagnostic trait "signatures", allowing us to discriminate among drivers by tracking shifts in community trait profiles. For a color version of this figure, see www.iste.co.uk/decocq/ecology.zip*

2.4.4. *Vectors of change – ordination*

Since the days of Whittaker and Curtis in the 1950s, ecologists have used multivariate methods to analyze plant community composition (Shipley 2021). Multivariate approaches are also useful for analyzing how communities respond to drivers of ecological change. For example, ordination allows us to visualize and compare patterns of community change. Analyzing both baseline and resurvey data together allows us to construct vectors reflecting how communities at each site have shifted in composition (Figure 2.4).

We can then assess how these changes differ among sites, regions or community types. Parallel vectors represent consistent responses. Converging vectors indicate biotic homogenization. Forests in Wisconsin show somewhat parallel changes but fragmented southern forests have clearly changed more. We can extend our use of ordination by examining how environmental factors (e.g. soil or landscape conditions) and plant traits covary (or "load") with ordination axes. Environmental factors that covary strongly with vectors of community (or trait) changes at particular sites are candidate likely drivers of those changes. Shifts in plant traits indicate how plant characteristics affect plant responses to drivers. These multivariate approaches deserve wider application.

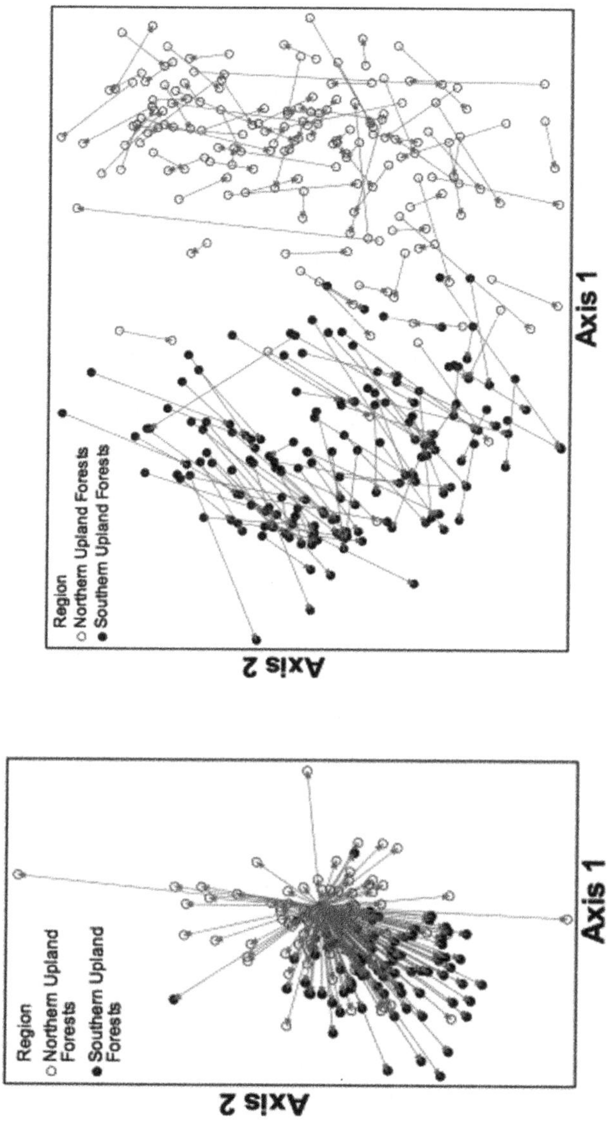

Figure 2.4. *Vectors of community change. Left: NMDS ordinations showing how the composition of plant communities in the fragmented deciduous forests of southern Wisconsin (solid points on left) and more continuous forests in northern Wisconsin (empty circles on right) changed over the past half-century. Right: placing the origins of change vectors together reveals that forests in both regions have shifted similarly, but that southern forests have changed more*

2.5. Conclusion

Ecological changes continue to grow in number and extent. All the environmental forces mentioned affect forest plant communities although their relative importance varies across regions and community types (Table 2.1). Forces like climate change are clearly growing in importance, while drivers like acid rain may be declining. As drivers act slowly and cumulatively, they exert both direct and indirect effects at multiple scales. This makes it hard to distinguish cause from effect ("passengers" vs. "drivers"). Applying new methods to growing datasets will allow us to better track the separate and combined impacts of different forces. It is particularly important to understand cumulative impacts and how drivers interact. In many areas, increases in clonal understory dominants like *Circaea, Carex* and *Parthenocissus* create "recalcitrant understory layers" that suppress the growth of tree seedlings and other species (Royo and Carson 2006). These layers often reflect herbivory and altered disturbance regimes. We ignore these synergistic effects at our peril. Surely, the effects of climate change will be compounded by habitat loss, preventing the adaptive migration of many species and restructuring communities (Radeloff et al. 2015).

The scarcity of historical baseline data remind us: today's surveys provide the bases for tomorrow's resurveys. The more we can standardize our field methods, which variables we track, and how we analyze our data, the more useful our results become. Satellite and aerial imagery, data sharing groups like sPlot, and standardized data from NEON (www.neonscience.org/) and the Forest Inventory and Analysis program (www.fia.fs.fed.us/) give us "big data" sets for studying ecological change. Synthesis centers like iDiv in Germany (www.idiv.de) are also enhancing our ability to compile, analyze and interpret ecological data by integrating vegetation plot surveys with climate, range and functional trait data (e.g. TRY, www.try-db.org). These efforts are yielding new insights of great value as we strive to understand, and adapt to, the forces driving ecological change.

2.6. References

Alstad, A.O., Damschen, E.I., Givnish, T.J., Harrington, J.A., Leach, M.K., Rogers, D.A., Waller, D.M. (2016). The pace of plant community change is accelerating in remnant prairies. *Sci. Adv.*, 2, e1500975.

Alverson, W.S., Waller, D.M., Solheim, S.L. (1988). Forests too deer: Edge effects in northern Wisconsin. *Conserv. Biol.*, 2, 348–358.

Alverson, W.S., Kuhlmann, W., Waller, D.M. (1994). *Wild Forests: Conservation Biology and Public Policy*. Island Press, Washington, DC.

Ash, J.D., Givnish, T.J., Waller, D.M. (2017). Tracking lags in historical plant species' shifts in relation to regional climate change. *Glob. Chang. Biolo.*, 23, 1305–1315.

Balgooyen, C.P. and Waller, D.M. (1995). The use of *Clintonia borealis* and other indicators to gauge impacts of white-tailed deer on plant communities in northern Wisconsin, USA. *Nat. Areas J.*, 15, 308–318.

Bartelheimer, M. and Poschlod, P. (2016). Functional characterizations of Ellenberg indicator values – A review on ecophysiological determinants. *Funct. Ecol.*, 30, 506–516.

Bernhardt-Römermann, M., Baeten, L., Craven, D., De Frenne, P., Hédl, R., Lenoir, J., Bert, D., Brunet, J., Chudomelová, M., Decocq, G. et al. (2015). Drivers of temporal changes in temperate forest plant diversity vary across spatial scales. *Glob. Chang. Biol.*, 21, 3726–3737.

Blossey, B. and Notzold, R. (1995). Evolution of increased competitive ability in invasive nonindigenous plants: A hypothesis. *J. Ecol.*, 83, 887–889.

Callaway, R.M. and Aschehoug, E.T. (2000). Invasive plants versus their new and old neighbors: A mechanism for exotic invasion. *Science*, 290, 521–523.

Chao, A., Chiu, C.H., Villéger, S., Sun, I.F., Thorn, S., Lin, Y.C., Chiang, J.M., Sherwin, W.B. (2019). An attribute-diversity approach to functional diversity, functional beta diversity, and related (dis)similarity measures. *Ecol. Monogr.*, 89, e01343.

Clark, C.M., Simkin, S.M., Allen, E.B., Bowman, W.D., Belnap, J., Brooks, M.L., Collins, S.L., Geiser, L.H., Gilliam, F.S., Jovan, S.E. et al. (2019). Potential vulnerability of 348 herbaceous species to atmospheric deposition of nitrogen and sulfur in the United States. *Nat. Plants*, 5, 697–705.

Côté, S.D., Rooney, T.P., Tremblay, J.-P., Dussault, C., Waller, D.M. (2004). Ecological impacts of deer overabundance. *Annu. Rev. Ecol. Evol. Syst.*, 35, 113–147.

Curtis, J.T. (1959). *The Vegetation of Wisconsin*. University of Wisconsin Press, Madison.

Dobson, A. and Blossey, B. (2015). Earthworm invasion, white-tailed deer and seedling establishment in deciduous forests of north-eastern North America. *J. Ecol.*, 103, 153–164.

Drayton, B. and Primack, R.B. (1996). Plant species lost in an isolated conservation area in metropolitan Boston from 1894 to 1993. *Conserv. Biol.*, 10, 30–39.

Gilliam, F.S., Welch, N.T., Phillips, A.H., Billmyer, J.H., Peterjohn, W.T., Fowler, Z.K., Walter, C.A., Burnham, M.B., May, J.D., Adams, M.B. et al. (2016). Year response of the herbaceous layer of a temperate hardwood forest to elevated nitrogen deposition. *Ecosphere*, 7, 1–16.

Heilmeier, H. (2019). Functional traits explaining plant responses to past and future climate changes. *Flora*, 254, 1–11.

Johnson, S.E., Amatangelo, K.L., Townsend, P.A., Waller, D.M. (2016). Large, connected floodplain forests prone to flooding best sustain plant diversity. *Ecology*, 97, 3019–3030.

Jost, L. (2007). Partitioning diversity into alpha and beta components. *Ecology*, 88, 2427–2439.

Kwiatkowska, A.J. (1994). Changes in the species richness, spatial pattern and species frequency associated with the decline of oak forest. *Vegetatio*, 112, 171–180.

Li, D. and Waller, D.M. (2015). Drivers of observed biotic homogenization in pine barrens of central Wisconsin. *Ecology*, 96, 1030–1041.

MacArthur, E.O. and Wilson, R.H. (1967). *The Theory of Island Biogeography*. Princeton University Press, Princeton.

Magnuson, J.J. (1990). Long-term ecological research and the invisible present: Uncovering the processes hidden because they occur slowly or because effects lag years behind causes. *Bioscience*, 40, 495–501.

Mayfield, M.M., Bonser, S.P., Morgan, J.W., Aubin, I., McNamara, S., Vesk, P.A. (2010). What does species richness tell us about functional trait diversity? Predictions and evidence for responses of species and functional trait diversity to land-use change. *Glob. Ecol. Biogeogr.*, 19, 423–431.

McGill, B.J., Enquist, B.J., Weiher, E., Westoby, M. (2006). Rebuilding community ecology from functional traits. *TREE*, 21, 178–185.

McKinney, M.L. and Lockwood, J.L. (1999). Biotic homogenization: A few winners replacing many losers in the next mass extinction. *Trends Ecol. Evol.*, 14, 450–453.

Parmesan, C. and Yohe, G. (2003). A globally coherent fingerprint of climate change impacts across natural systems. *Nature*, 421, 37–42.

Radeloff, V.C., Williams, J.W., Bateman, B.L., Burke, K.D., Carter, S.K., Childress, E.S., Cromwell, K.J., Gratton, C., Hasley, A.O., Kraemer, B.M. et al. (2015). The rise of novelty in ecosystems. *Ecol. Appl.*, 25, 2051–2068.

Rogers, D.A., Rooney, T.P., Olson, D., Waller, D.M. (2008). Shifts in southern Wisconsin forest canopy and understory richness, composition, and heterogentity. *Ecology*, 89, 2482–2492.

Rolhauser, A.G., Waller, D.M., Tucker, C.M. (2021). Complex trait–environment relationships underlie the structure of forest plant communities. *J. Ecol.*, 109, 3794–3806.

Rooney, T.P. and Dress, W.J. (1997). Patterns of plant diversity in overbrowsed primary and mature secondary hemlock northern hardwood forest stands. *J. Torrey Bot. Soc.*, 124, 43–51.

Rooney, T.P., Wiegmann, S.M., Rogers, D.A., Waller, D.M. (2004). Biotic impoverishment and homogenization in unfragmented forest understory communities. *Conserv. Biol.*, 18, 787–798.

Royo, A.A. and Carson, W.P. (2006). On the formation of dense understory layers in forests worldwide: Consequences and implications for forest dynamics, biodiversity, and succession. *Can. J. For. Res.*, 36, 1345–1362.

Shen, X., Bourg, N.A., McShea, W.J., Turner, B.L., Rooney, T., Dress, W., Rooney, T., Wiegmann, S., Rogers, D.A., Waller, D.M. et al. (2016). Long-term effects of white-tailed deer exclusion on the invasion of exotic plants: A case study in a Mid-Atlantic temperate forest. *PLoS One*, 11, e0151825.

Shipley, B. (2021). *Ordination Methods for Biologists: A Non-Mathematical Introduction Using R*. BS Publishing, Sherbrooke.

Simkin, S.M., Allen, E.B., Bowman, W.D., Clark, C.M., Belnap, J., Brooks, M.L., Cade, B.S., Collins, S.L., Geiser, L.H., Gilliam, F.S. et al. (2016). Conditional vulnerability of plant diversity to atmospheric nitrogen deposition across the USA. *Proc. Natl. Acad. Sci. USA*, 113, 4086–4091.

Staude, I.R., Waller, D.M., Bernhardt-Römermann, M., Bjorkman, A.D., Brunet, J., De Frenne, P., Hédl, R., Jandt, U., Lenoir, J., Máliš, F. et al. (2020). Replacements of small- by large-ranged species scale up to diversity loss in Europe's temperate forest biome. *Nat. Ecol. Evol.*, 4, 802–808.

Stevens, C.J., David, T.I., Storkey, J. (2018). Atmospheric nitrogen deposition in terrestrial ecosystems: Its impact on plant communities and consequences across trophic levels. *Funct. Ecol.*, 32, 1757–1769.

Suding, K.N., Collins, S.L., Gough, L., Clark, C., Cleland, E.E., Gross, K.L., Milchunas, D.G., Pennings. S. (2005). Functional- and abundance-based mechanisms explain diversity loss due to N fertilization. *Proc. Natl. Acad. Sci. USA*, 102, 4387–4392.

Thomas, H.J.D., Bjorkman, A.D., Myers-Smith, I.H., Elmendorf, S.C., Kattge, J., Diaz, S., Vellend, M., Blok, D., Cornelissen, J.H.C., Forbes, B.C. et al. (2020). Global plant trait relationships extend to the climatic extremes of the tundra biome. *Nat. Commun.*, 111(11), 1–12.

Vellend, M., Verheyen, K., Jacquemyn, M.H., Kold, A., Peterken, G., Hermy, M. (2006). Extinction debt of forest plants persists for more than a century following habitat fragmentation. *Ecology*, 87, 542–548.

Verheyen, K., Baeten, L., De Frenne, P., Bernhardt-Römermann, M., Brunet, J., Cornelis, J., Decocq, G., Dierschke, H., Eriksson, O., Hédl, R. et al. (2012). Driving factors behind the eutrophication signal in understorey plant communities of deciduous temperate forests. *J. Ecol.*, 100, 352–365.

Verheyen, K., De Frenne, P., Baeten, L., Waller, D.M., Hédl, R., Perring, M.P., Blondeel, H., Brunet, J., Chudomelová, M., Decocq, G. et al. (2017). Combining biodiversity resurveys across regions to advance global change research. *Bioscience*, 67, 73–83.

Waller, D.M. and Alverson, W.S. (1997). The white-tailed deer: A keystone herbivore. *Wildl. Soc. Bull.*, 25, 217–226.

Waller, D.M. and Reo, N.J. (2018). First stewards: Ecological outcomes of forest and wildlife stewardship by indigenous peoples of Wisconsin, USA. *Ecol. & Soc.*, 23 [Online]. Available at: https://doi.org/10.5751/ES-09865-230145.

Waller, D.M., Amatangelo, K., Johnson, S., Rogers, D.A. (2012). Wisconsin vegetation database – Plant community survey and resurvey data from the Wisconsin plant ecology laboratory. *Biodivers. Ecol.*, 4, 255–264.

Waller, D.M., Mudrak, E.L., Rogers, D.A. (2017). Do metacommunity mass effects predict changes in species incidence and abundance? *Ecography*, 41, 11–23.

WICCI (2010). Wisconsin initiative on climate change impacts [Online]. Available at: http://www.wicci.wisc.edu/.

3

Getting the Right Answer Can Take a While: Long-term Ecological Field Studies as Historical Ecology

Frank S. GILLIAM
University of West Florida, Pensacola, USA

3.1. Introduction

Implicit in any discussion of the essential importance of historical ecology as an ecological discipline is the dimension of temporal scale (Szabó and Hédl 2011). Indeed, among the numerous challenges in ecological studies is that of addressing both temporal and spatial scales, which Levin (1992) identified as a central problem in ecology, one that unifies such otherwise-disparate subdisciplines as population ecology and ecosystem ecology, as well as basic and applied ecology. For example, essential ecosystem drivers and related distribution of vegetation operate on spatial scales from just over 0 to >10,000 km and temporal scales from hours to millennia. Ecosystem drivers would include weather, climate variability/climate change, fire and land-use change, whereas vegetation ranges from local patches to global biomes (Figure 3.1, top). It is not coincidental that the spatial/temporal scales of historical ecology parallel those of both the variety of ecological subdisciplines and the study of environmental drivers and vegetation (Figure 3.1, bottom).

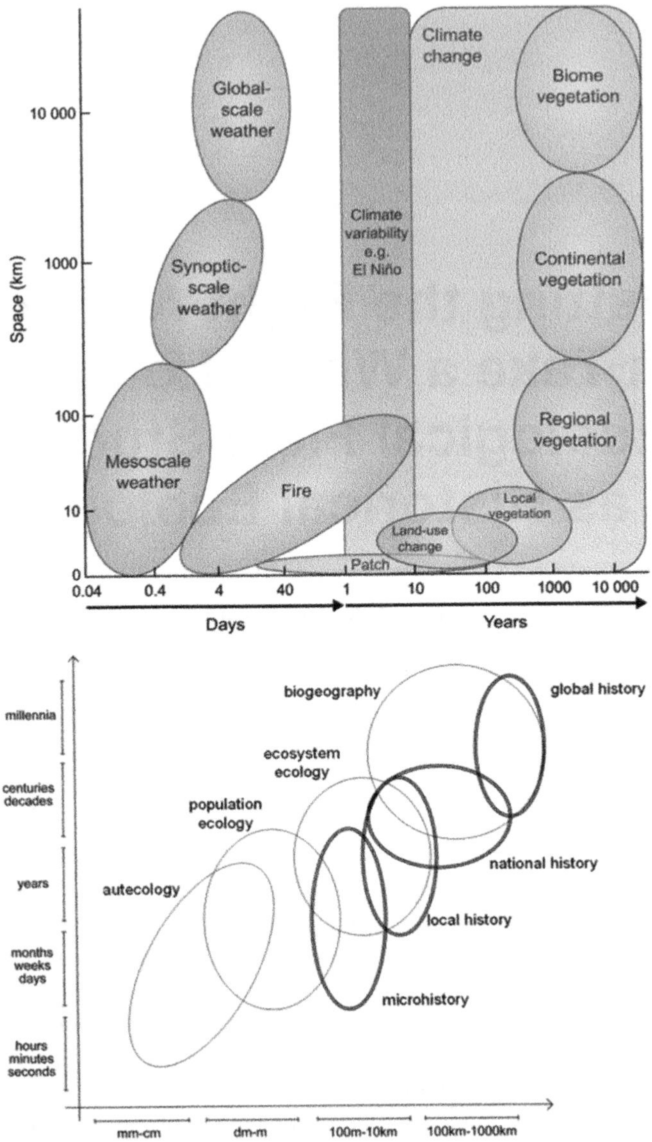

Figure 3.1. *Spatial and temporal scales of (top) essential ecosystem drivers (weather, climate variability and climate change, fire and land-use change) and related distribution of vegetation. Figure reprinted from Gilliam (2016), used with permission, and (bottom) subdisciplines of the study of ecology and of the study of history. Figure reprinted from Szabó and Hédl (2011) with permission. For a color version of this figure, see www.iste.co.uk/decocq/ecology.zip*

It is imperative to study ecological systems at the appropriate temporal scale, with most field-based endeavors studying phenomena that operate under longer periods. Such studies occasionally yield new questions to address when novel ecological patterns and processes emerge through their extended duration. Perhaps, the best example of this is the Park Grass Experiment, currently the longest ongoing field experiment in the world (Richardson 1938). It was initiated in 1856 at Rothamsted in Hertfordshire, England, by John B. Lawes and Joseph H. Gilbert, who wanted to address a very pragmatic concern for English dairy farmers – how to improve hay yield via the addition of a variety of fertilizers (Silvertown et al. 2006). Although it took relatively little time to answer that basic question, maintaining the experiment over a far longer period allowed researchers to address fundamental ecological processes never considered by Lawes and Gilbert, including effects of plant biomass and excess nitrogen on plant diversity, trophic dynamics of an herb-dominated ecosystem and genetic drift in plant populations (Silvertown et al. 2006). In addition, such extended research eventually led to an awareness of the importance of long-term studies in general (Likens 1989), including whole research programs, such as the Long-Term Ecological Research and Long-Term Research in Environmental Biology programs of the National Science Foundation in the United States.

The purpose of this chapter is to examine the importance of maintaining long-term studies to generate and test hypotheses regarding the ecology of terrestrial ecosystems, in this case, a hardwood forest ecosystem of the central Appalachian region of the United States, namely, the Fernow Experimental Forest (FEF), West Virginia, U.S.A. As it is not uncommon for papers to be published based on three to five years of field data, of particular interest is how extending field sampling to over nearly 30 years has, in many cases, led to conclusions quite different from those based on shorter time periods. In addition to considerable work carried out by scientists at the USDA Forest Service Timber and Watershed Laboratory, Parsons, West Virginia, where facilities associated with FEF are located, FEF has been the site of numerous cooperative studies by scientists from colleges and universities primarily in the eastern U.S. Rather than attempt to review all such studies, this chapter covers primarily only work carried out by the Weeds and Dirt Laboratory at Marshall University, Huntington, West Virginia.

3.2. Fernow Experimental Forest

3.2.1. *Background*

Forest ecology at FEF has a long and productive research history (Adams et al. 2006). Initial work was on silvicultural practices (Smith and Miller 1987), with the collection of hydrochemical data for the long-term reference watershed (WS4) beginning in 1960. A specific focus on the effects of acid deposition was initiated via the Watershed Acidification Study (WAS), which began as a now-terminated pilot study in 1987 on a watershed adjacent to FEF. In 1989, it was established on FEF watersheds and is currently ongoing, although treatments ceased in 2019. A distinctive feature of the WAS is that has involved a whole-watershed application of simulated acidic deposition – as $(NH_4)_2SO_4$ – via three aerial additions of year, representing total N addition of 35 kg N ha^{-1} yr^{-1}. This has been applied in powder form by airplane (Adams et al. 2006).

3.2.2. *Site description*

Fernow occupies ~1,900 ha of the Allegheny Mountain section of the unglaciated Allegheny Plateau near Parsons, West Virginia (39°03'16"N, 79°41'0"W). Precipitation at FEF averages approximately 1,430 mm yr^{-1}, generally higher during the growing season and increasing with elevation. As of the initiation of the WAS, concentrations of acidic species in wet deposition (H^+, $SO_4^=$ and NO_3^-) were among the highest in North America. Watershed soils are coarse-textured Inceptisols (loamy-skeletal, mixed mesic Typic Dystrochrept) of the Berks and Calvin series sandy loams derived from sandstone (Adams et al. 2006).

Dominant tree species on FEF watersheds vary with stand age. Early-successional species such as black birch (*Betula lenta* L.), black cherry (*Prunus serotina* Ehrh.) and yellow-poplar (*Liriodendron tulipifera* L.) are dominant in young stands, whereas late-successional species, such as sugar maple (*Acer saccharum* Marshall) and northern red oak (*Q. rubra* L.), are dominant in mature stands. Dominant herbaceous layer species at the beginning of the study varied less with stand age, including stinging nettle (*Laportea canadensis* (L.) Wedd.), violets (*Viola* spp.) and several ferns (Gilliam et al. 2006).

3.2.3. *Field design*

Three watersheds are used for ongoing research as part of the WAS (Figure 3.2). WS4 supports >100-year-old even-aged stands, serving as the long-time reference watershed at FEF. WS7 supports an approximately 40-year-old even-age stand, whereas WS3 supports an approximately 40-year-old even-age stand and serves as the treatment watershed, whereas WS4 and WS7 are the controls. WS3 has received three aerial applications of $(NH_4)_2SO_4$ yr^{-1}, beginning in 1989. March (or sometimes April) and November applications represent approximately 7.1 kg N ha^{-1}; July applications are approximately 21.2 kg N ha^{-1}. The total amount of N deposited on

WS3 (application plus atmospheric deposition) is approximately 54 kg N ha^{-1} yr^{-1}, or about three times ambient inputs (Adams et al. 2006).

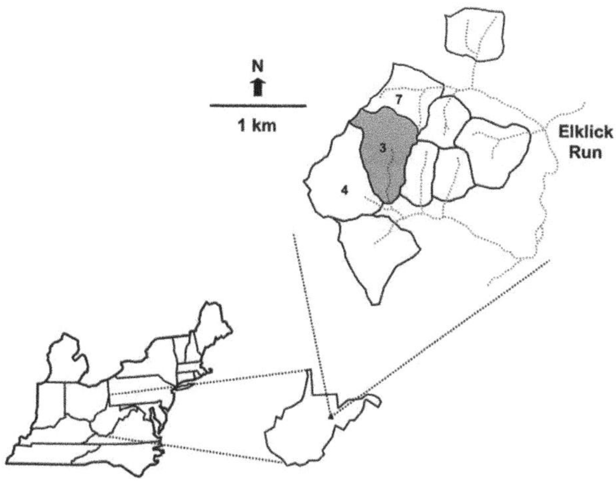

Figure 3.2. *Experimental watersheds of the Fernow Experimental Forest, West Virginia. WS3, WS4 and WS7 are used as part of the ongoing Watershed Acidification Study*

In 1991, 15 permanent vegetation plots were established in each of four watersheds of contrasting stand age at FEF, with a particular focus on interactions between the forest canopy and the herbaceous stratum, as well as plant–soil interactions, and how these interactions might vary with forest succession following disturbance. A sub-set of seven of these original plots in each of three watersheds – WS3, WS4 and WS7 – were selected in 1993 for monthly in situ incubation of soil to examine spatial and temporal patterns of net N mineralization and nitrification. This represented the first work at FEF to specifically address a phenomenon that until that time had been described more in Europe and experimental studies in the northeastern United States, *N saturation*, which arises from atmospheric inputs of N exceeding biological N demand (Aber et al. 1998). This typically results in the loss of NO_3^- in streams and is commonly accompanied by the loss of nutrients (e.g. Ca^{++} and Mg^{++}) (the mobile anion effect) that are essential to plant growth and forest health. Some earlier studies published on N saturation in the US (e.g. Stoddard 1994; Gilliam et al. 1996; Peterjohn et al. 1996) identified watersheds at FEF to be among the more N-saturated sites in North America. Symptoms included increased mobility and leaching of Ca^{++} and Mg^{++} as being linked to increased N deposition, along with enhanced nitrification and movement of NO_3^- and decreases in growth rates of dominant tree species (Peterjohn et al. 1996; Christ et al. 2002; May et al. 2005; Adams et al. 2006). Other problems associated with N saturation include the increased production of the greenhouse gas, N_2O (Peterjohn et al. 1998). Further work at FEF has also suggested that N saturation has led to phosphorous limitation in several watersheds (Gress et al. 2007).

3.3. Long-term studies at Fernow Experimental Forest, West Virginia

There are several examples from FEF wherein initial conclusions made after a shorter period of investigation were contradicted by extending sampling over greater time periods. In general, the original conclusion was a lack of effect of experimental additions on treatment WS3, based on three to five years of treatment, often with a focus on the N component of these additions. It should be noted, however, that the numerous studies at other forest sites had, at that time, shown significant treatment effects after a similar time period, especially regarding the response of the herb layer (Hurd et al. 1998; Lu et al. 2010; Chapman et al. 2016), so the published lack of treatment effects at FEF was notable in themselves. It should also be noted that some of these studies focus on the experimental treatment as simulated acid deposition, whereas others specifically focused on the N component of the treatment.

3.3.1. *Effects of acidification on soil fertility and herb layer cover and foliar nutrients*

Gilliam et al. (1994) examined the effects of experimental acidification on soil fertility and several characteristics of the herb layer using all three watersheds of the WAS following a three-year treatment period for WS3. They found essentially no significant differences in any of the measured soil variables – pH, organic matter, cation exchange capacity, extractable nutrients – among watersheds, as well as no differences in herb cover, biomass and foliar nutrients. This led to the conclusion that three years of whole-watershed acidification had minimal, if any, effects on nutrient availability – a surprising result at the time.

As a follow-up to Gilliam et al. (1994), in July 2015 – 24 years after the original sampling in 1991 – Gilliam et al. (2020) resampled mineral soil from the 7 of the original 15 plots in each of the three WAS watersheds. Although the focus was on temporal shifts in effects of experimental acidification on soil fertility, they also reported on extensive stream chemistry data, including concentrations of H^+, NO_3^-, base cations and electrical conductivity (Figure 3.3). Gilliam et al. (2020) yielded sharply contrasting results from Gilliam et al. (1994), wherein virtually all measured variables exhibited significant treatment effects. In general, pH was lower, exchangeable acidity and aluminum were higher, nutrient (base) cations were lower, and extractable N was higher following a quarter century of experimental acidification.

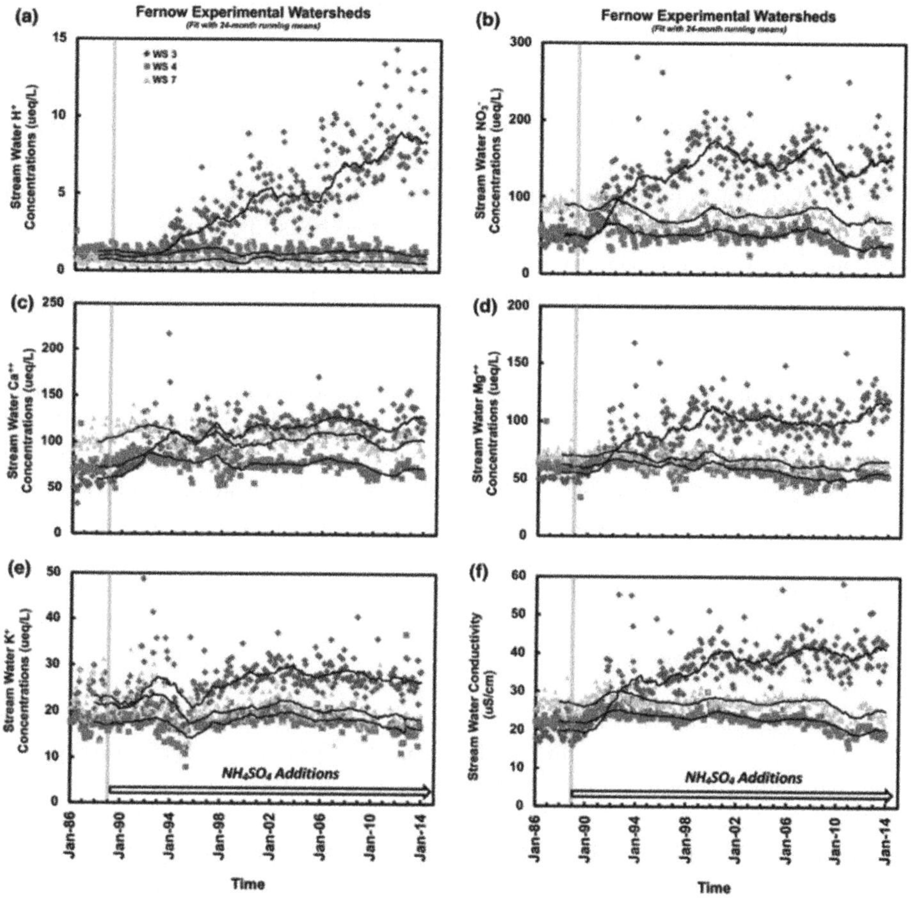

Figure 3.3. *Volume-weighted mean monthly concentrations of major ions in stream water and electrical conductivity from 1989 to 2014 for three small watersheds at the Fernow Experimental Forest in West Virginia. Trend lines are 24-mo running means. Vertical bars indicate when fertilizer applications began. Figure reprinted from Gilliam et al. (2020) with permission. For a color version of this figure, see www.iste.co.uk/decocq/ecology.zip*

3.3.2. *Effects of N addition on soil N dynamics*

In 1993, the same subset of seven plots of the original 15 were identified for in situ ("buried bag") incubations for measurements of net N mineralization and nitrification in surface mineral soil in each of the three WAS watersheds. Briefly, this method involved sampling of mineral soil to a 5-cm depth at five locations in each plot and combining/mixing soil to yield a single composite sample, which was placed in two sterile polyethylene bags, one of which was buried at a 5-cm depth and the other was taken to the lab for immediate extraction and analysis for NH_4^+ and NO_3^-. This was repeated monthly during the growing season on all 21 plots from 1993 to 1995, with each collection comprised of two bags: the buried bag from the previous month and the bag from the current month. Differences in extractable N between a non-buried bag and its buried equivalent were used to determine rates of net N mineralization and nitrification.

Gilliam et al. (1996) reported no effects of experimental N additions on N mineralization and nitrification for the first year of study, but found significant seasonal variation. Gilliam et al. (2001) extended the sample period for the three-year period of 1993–1995, finding significant, but transient, monthly effects on net nitrification, wherein rates were significantly higher on WS3 for months that immediately followed N additions. They also took advantage of the extended sampling period to use multiple regression with ambient environmental variables, concluding that net nitrification was directly influenced by mean ambient temperature and soil moisture.

Following subsequent sampling in 2003 and 2005, the yearly sampling on these plots was initiated in 2007 with funding from the NSF Long Term Research in Environmental Biology (NSF LTREB) program to West Virginia University (William T. Peterjohn, principal investigator). This work is currently ongoing. Gilliam et al. (2018) reported on numerous aspects of N dynamics from the 21-year period from 1993 to 2014, demonstrating that there had been a transient effect of N additions on the annual mean net nitrification in 1995, after which all watersheds exhibited similar declines to 2005 and are currently increasing. They found that the best predictor of net nitrification was a metric of ambient temperature – number of degree days <19°C – a significant pattern did not emerge until after the longer period of study (Figure 3.4).

3.3.3. *Effects of N addition on herb layer composition and diversity*

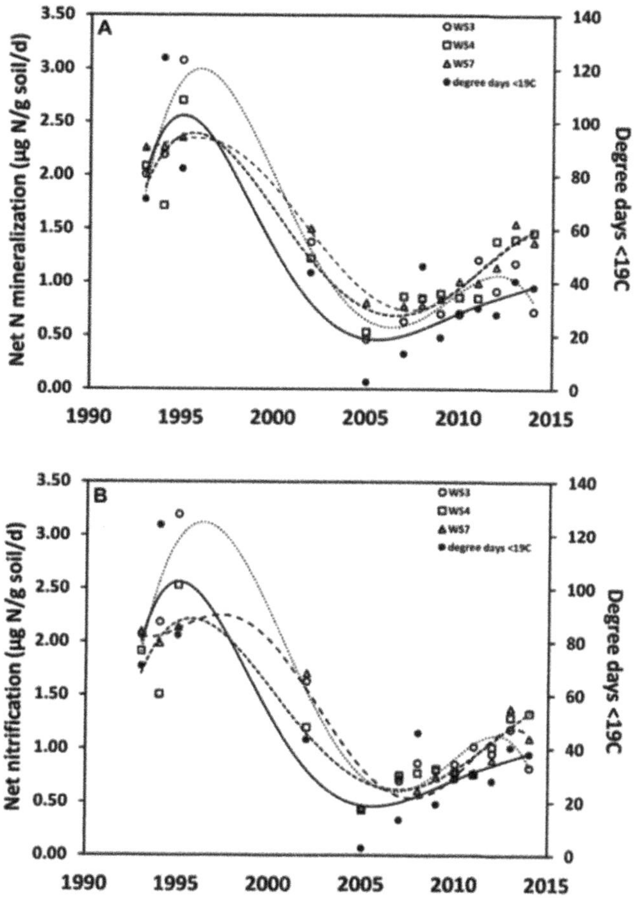

Figure 3.4. *Mean annual growing season net N mineralization (A) and net nitrification (B) in mineral soil for study watersheds at Fernow Experimental Forest, West Virginia, 1993–2014. Annual degree days < 19°C are shown. Fitted curves are fifth-order polynomials for all variables to visually characterize temporal trends. Degree days: closed circles/solid curve,* r^2 *= 0.82; WS3: open circles/fine dashes,* r^2 *= 0.94 and 0.98 for net N mineralization and net nitrification, respectively; WS4: open squares/intermediate dashes,* r^2 *= 0.89 and 0.89; WS7: open triangles/wide dashes,* r^2 *= 0.96 and 0.95. Figure reprinted from Gilliam et al. (2018) with permission*

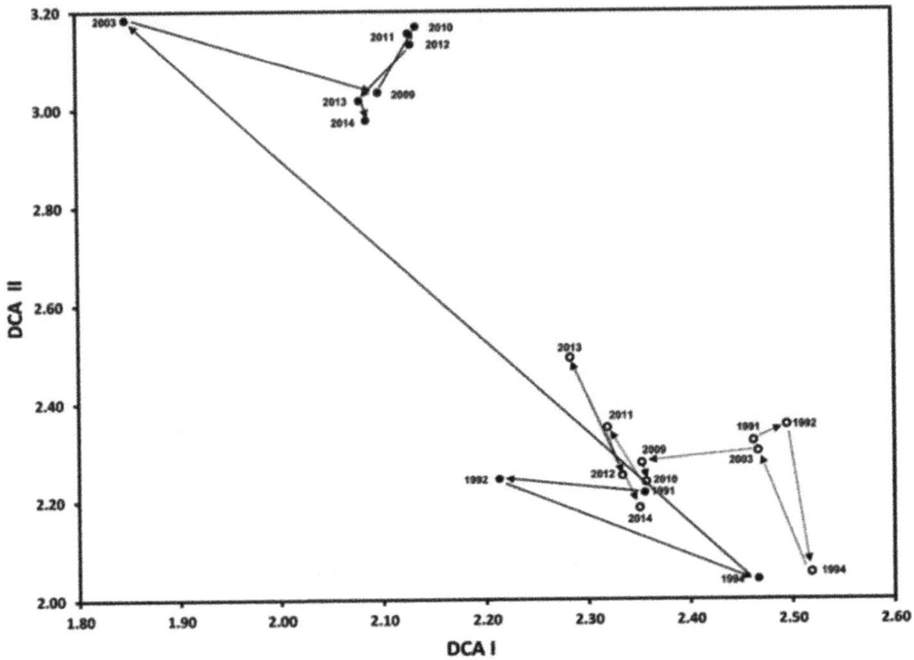

Figure 3.5. *Detrended correspondence analysis (DCA) for herbaceous layer communities from 1991 to 2014 on reference WS4 (open symbols/dashed line) and N-treated WS3 (solid symbols/solid line). Values shown are centroids (two-dimensional means) for each year per watershed. Figure reprinted from Gilliam et al. (2016) with permission*

The initial conclusions regarding the effects of experimental acidification (Gilliam et al. 1994) were based on a single year of sampling across all watersheds (1991). Herb layer sampling was continued on WS3 and WS4 in 1992 and 1994, representing a treatment period of three to six years on WS3, a time period during which other studies had reported significant responses to experimentally added N (e.g. Hurd et al. 1998; Lu et al. 2010). Thus, it was surprising that, in synthesizing all data through 1994, Gilliam et al. (2006) reported no effects of added N on herb cover, species richness, evenness and diversity at FEF. In addition, there was minimal variation in species composition, both between treatment and reference watershed and through time, from 1991 to 1994. They concluded that the lack of treatment effect was from the high ambient N deposition at FEF, such that the amount of added N (35 kg N/ha/yr) was a smaller percentage relative to ambient deposition compared to other studies.

Herb layer sampling was repeated on a subset of 7 of the original 15 plots on WS3 and WS4 in 2004. Again, as a result of the NSF LTREB funding, annual sampling was initiated in 2009 and is currently ongoing on WS3 and WS4 only. Gilliam et al. (2016) reported results up to 2014, a total of 10 yr of data taken over a 23-year period. They found significant and notable effects of excess N on all aspects of the herb layer community, including increases in cover and decreases in species richness, evenness and diversity (Figure 3.5). There were also profound changes in composition, with cover of *Rubus allegheniensis* – once a minor component on all watersheds – increasing >10-fold on N-treated WS3. The response of *R. allegheniensis* to N additions to WS3 has been confirmed by other studies at FEF (Walter et al. 2016, 2017).

3.3.4. *The N homogeneity hypothesis*

One of the more salient outcomes of long-term research at FEF was the development – and eventual testing – of a hypothesis that addresses the effects of excess N on forest herb communities, *the N homogeneity hypothesis* (Gilliam 2006). The hypothesis predicts the loss of herb layer diversity from anthropogenically increased N deposition via the following: (1) herb cover will initially increase (i.e. a fertilizer effect), (2) N-mediated increases in herb cover simultaneously enhances the growth of nitrophilous species and elimination of N-efficient species, (3) both species richness and evenness will decline (i.e. from increased dominance of only few nitrophilic species), contributing to declines in diversity, and (4) excess N will increase the spatial homogeneity of N availability, contributing to increased sample homogeneity of the herb layer (Gilliam et al. 2016). A key facet of the mechanism behind this response is the importance of spatial heterogeneity of soil resources in maintaining high diversity in plant communities (Hutchings et al. 2003).

Initial observations for the hypothesis began as early as Gilliam et al. (2001), who reported lower spatial variation (measured as coefficients of variation of watershed-scale means) in net nitrification and extractable NO_3^- pools in mineral soil of treatment WS3 relative reference watersheds WS7 and WS4. It was originally articulated as a predictive hypothesis in a review of responses of the herb layer of forests of North America and Europe to excess N deposition (Gilliam 2006). Although this hypothesis has received wide support in the literature (Cholewińska et al. 2020), it was only able to be tested at FEF by extending the original sampling to the current long-term period, including a more extensive one-time spatial sampling of 100 plots in each of treatment WS3 and reference WS7 in 2011 (Figure 3.6). As reported in Gilliam et al. (2016), the spatial homogeneity of net nitrification was quite high on WS3, relative to WS7. Using the ongoing sampling from permanent plots on all three watersheds, Gilliam et al. (2018) demonstrated N-mediated increases in spatial homogeneity of net nitrification over time on WS3 up to a quarter century following the initiation of treatments.

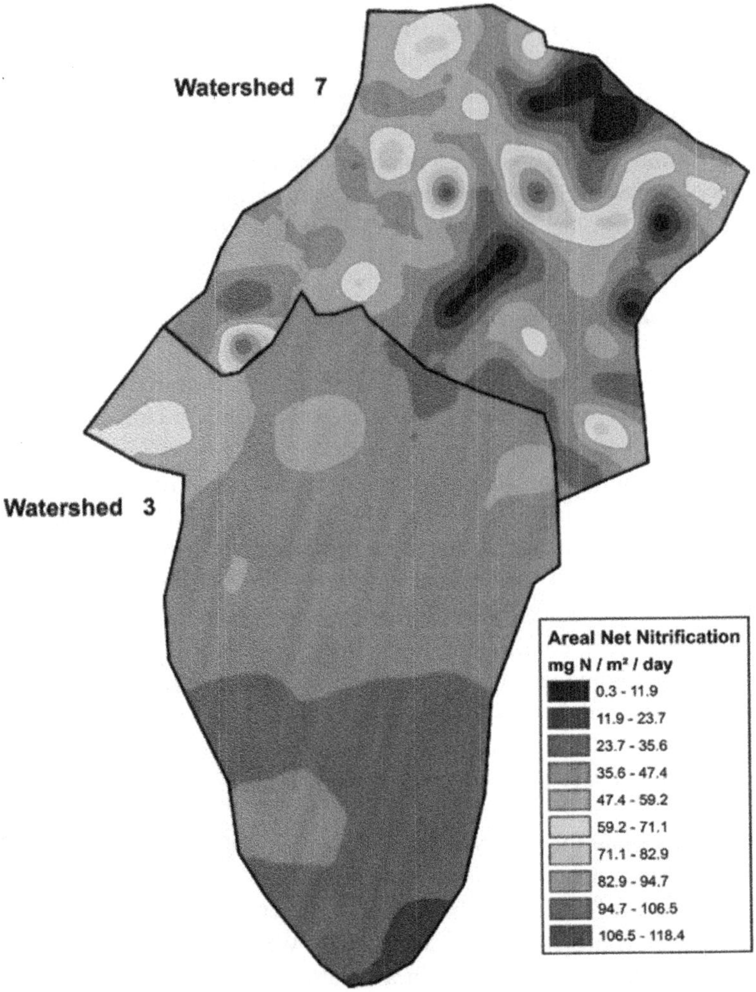

Figure 3.6. *Spatial patterns of the potential net nitrification measured in 2011 using surface soils (top 5 cm of mineral soil) collected from the reference (WS7) and N-treated (WS3) watersheds at the Fernow Experimental Forest, West Virginia. Figure reprinted from Gilliam et al. (2016) with permission. For a color version of this figure, see www.iste.co.uk/decocq/ecology.zip*

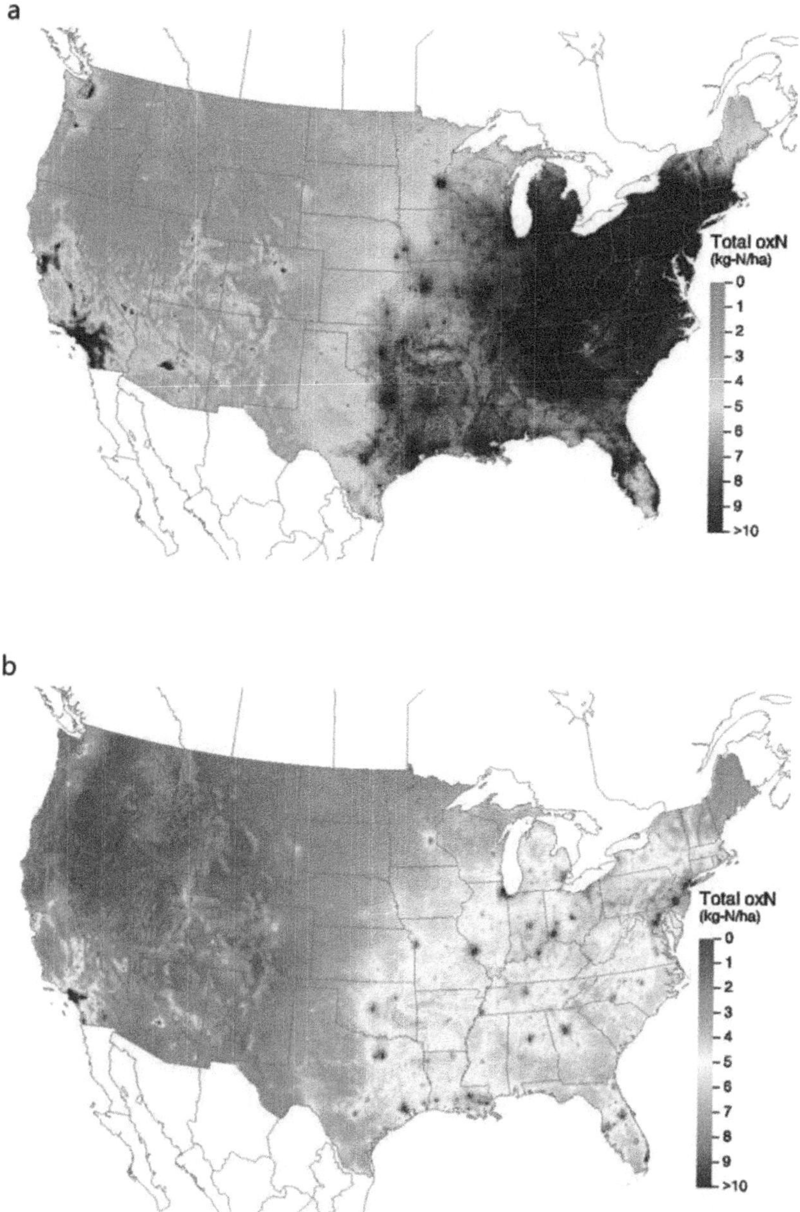

Figure 3.7. *Total deposition of NOx for 2000 (a) and 2014 (b). Figure reprinted from Gilliam et al. (2019) with permission. For a color version of this figure, see www.iste.co.uk/decocq/ecology.zip*

3.3.5. *A look at the future: declines in the atmospheric deposition of N*

During the decade preceding the initiation of the WAS, the wet deposition of N at FEF averaged 7.3 kg N/ha/yr, whereas mean deposition during the most recent decade has been 4.2 kg N/ha/yr, a decline of >40%. This decline has been reported throughout North America (Du 2016; Lloret and Valiela 2016; Groffman et al. 2018; Gilliam et al. 2019) and Europe (Schmitz et al. 2019), the result of regulatory legislation in many parts of the world, such as the Clean Air Act in the U.S. (Du et al. 2019). Because these global clean-air policies primarily target emissions of N associated with energy combustion, declines in atmospheric N deposition are far more prominent for NO_3^- than for NH_4^+ (Figure 3.7).

Such a trend does not call into question the relevance of long-term studies at FEF or similar studies elsewhere. Rather, by maintaining such studies following cessation of the N treatment on WS3, this work will become uniquely positioned to test hypotheses regarding the legacies of chronic N deposition and N saturation (Groffman et al. 2018; Gilliam et al. 2019; Schmitz et al. 2019). Groffman et al. (2018) have suggested that impacted forests will undergo N oligotrophication as increased carbon (C) flow from the atmosphere stimulates microbial N mobilization in forest soil, decreasing available N for plants. Gilliam et al. (2019) proposed a hysteretic model for decreased N future, wherein there will be widely varying time lags in the recovery of soil acidification, plant biodiversity, soil microbial communities, C and N cycling, and surface water chemistry towards pre-N impact conditions. They concluded that, although there is a great deal of uncertainty regarding forest response to future environmental change, it is quite certain that ambient CO_2 and temperature following any return to pre-impact status will not be the same as they were prior to past increases in N deposition.

3.4. Conclusion

Historical ecology is inexorably linked with temporal scale, as it focuses on past ecosystems and regards humans as one of numerous factors influencing these systems (Szabó and Hédl 2011; Szabó 2015). Indeed, it can be argued that implicit in the very study of ecology is a historical dimension (Figure 3.1), as ecologists often seek to predict future responses of ecological systems, further predicting spatial variation in their structure and function. This has become increasingly relevant as ecologists consider ecosystem response to global change (Gilliam 2016). Three decades of research at the Fernow Experimental Forest as part of the Watershed Acidification Study underscores the essentiality of maintaining field studies over long-term periods of time to address ecological processes, especially in the context of global change.

3.5. References

Aber, J., McDowell, W., Nadelhoffer, K., Magill, A., Berntson, G., Kamakea, M., McNulty, S., Currie, W., Rustad, L., Fernandez, I. (1998). Nitrogen saturation in temperate forest ecosystems – Hypotheses revisited. *BioScience*, 48, 921–934.

Adams, M.B., De Walle, D.R., Hom, J. (2006). *The Fernow Watershed Acidification Study*. Springer, Dordrecht, The Netherlands.

Chapman, S.K., Devine, K.A., Curran, C., Jones, R.O., Gilliam, F.S. (2016). Impacts of soil nitrogen and carbon additions on forest understory communities with a high-deposition history. *Ecosystems*, 19, 142–154.

Cholewińska, O., Adamowski, W., Jaroszewicz, B. (2020). Homogenization of temperate mixed deciduous forests in Białowieża Forest: Similar communities are becoming more similar. *Forests*, 11, 545.

Christ, M.J., Peterjohn, W.T., Cumming, J.R., Adams, M.B. (2002). Nitrification potentials and landscape, soil and vegetation characteristics in two Central Appalachian watersheds differing in NO_3^- export. *For. Ecol. Manage.*, 159, 145–158.

Du, E. (2016). Rise and fall of nitrogen deposition in the United States. *Proc. Natl. Acad. Sci. USA*, 113, E3594–E3595.

Du, E., Fenn, M.E., De Vries, W. (2019). Atmospheric nitrogen deposition to global forests: Status, impacts and management option. *Environ. Pollut.*, 250, 1044–1048.

Gilliam, F.S. (2006). Response of the herbaceous layer of forest ecosystems to excess nitrogen deposition. *J. Ecol.*, 94, 1176–1191.

Gilliam, F.S. (2016). Forest ecosystems of temperate climatic regions: From ancient use to climate change. *New Phytol.*, 212, 871–887.

Gilliam, F.S., Turrill, N.L., Aulick, S.D., Evans, D.K., Adams, M.B. (1994). Herbaceous layer and soil response to experimental acidification in a central Appalachian hardwood forest. *J. Environ. Qual.*, 23, 835–844.

Gilliam, F.S., Yurish, B.M., Adams, M.B. (2001). Temporal and spatial variation of nitrogen transformations in nitrogen-saturated soils of a central Appalachian hardwood forest. *Can. J. For. Res.*, 31, 1768–1785.

Gilliam, F.S., Hockenberry, A.W., Adams, M.B. (2006). Effects of atmospheric nitrogen deposition on the herbaceous layer of a central Appalachian hardwood forest. *J. Torrey Bot. Soc.*, 133, 240–254.

Gilliam, F.S., Welch, N.T., Phillips, A.H., Billmyer, J.H., Peterjohn, W.T., Fowler, Z.K., Walter, C.A., Burnham, M.B., May, J.D., Adams, M.B. (2016). Twenty-five year response of the herbaceous layer of a temperate hardwood forest to elevated nitrogen deposition. *Ecosphere*, 7, e01250. 10.1002/ecs2.1250.

Gilliam, F.S., Walter, C.A., Adams, M.B., Peterjohn, W.T. (2018). Nitrogen (N) dynamics in the mineral soil of a central Appalachian hardwood forest during a quarter century of whole-watershed N additions. *Ecosystems*, 21, 1489–1504.

Gilliam, F.S., Burns, D.A., Driscoll, C.T., Frey, S.D., Lovett, G.M., Watmough, S.A. (2019). Decreased atmospheric nitrogen deposition in eastern North America: Predicted responses of forest ecosystems. *Environ. Pollut.*, 244, 560–574.

Gilliam, F.S., Adams, M.B., Peterjohn, W.T. (2020). Response of soil fertility to 25 years of experimental acidification in a temperate hardwood forest. *J. Environ. Qual.*, 49, 961–972.

Gress, S.E., Nichols, T.D., Northcraft, C.C., Peterjohn, W.T. (2007). Nutrient limitation in soils exhibiting differing nitrogen availabilities: What lies beyond nitrogen saturation? *Ecology*, 88, 119–130.

Groffman, P.M., Driscoll, C.T., Durán, J., Campbell, J.L., Christenson, L.M., Fahey, T.J., Fisk, M.C., Fuss, C., Likens, G.E., Lovett, G. et al. (2018). Nitrogen oligotrophication in northern hardwood forests. *Biogeochemistry*, 141, 523–539.

Hurd, T.M., Brach A.R., Raynal D.J. (1998). Response of understory vegetation of Adirondack forests to nitrogen additions. *Can. J. For. Res.*, 28, 799–807.

Hutchings, M.J., John, E.A., Wijesinghe, D.K. (2003). Toward understanding the consequences of soil heterogeneity for plant populations and communities. *Ecology*, 84, 2322–2334.

Levin, S.A. (1992). The problem of pattern and scale in ecology. *Ecology*, 73, 1943–1967.

Likens, G.E. (1989). *Long-Term Studies in Ecology: Approaches and Alternatives.* Springer-Verlag, New York.

Lloret, J. and Valiela, I. (2016). Unprecedented decrease in deposition of nitrogen oxides over North America: The relative effects of emission controls and prevailing airmass trajectories. *Biogeochemistry*, 129, 165–180.

Lu, X., Mo, J., Gilliam, F.S., Zhou, G., Fang, Y. (2010). Effects of experimental nitrogen deposition on plant diversity in an old-growth tropical forest. *Glob. Change Biol.*, 16, 2688–2700.

May, J.D., Burdette, E., Gilliam, F.S., Adams, M.B. (2005). Interspecific divergence in foliar nutrient dynamics and stem growth in a temperate forest in response to chronic nitrogen inputs. *Can. J. For. Res.*, 35, 1023–1030.

Peterjohn, W.T., Adams, M.B., Gilliam, F.S. (1996). Symptoms of nitrogen saturation in two central Appalachian hardwood forests. *Biogeochemistry*, 35, 507–522.

Peterjohn, W.T., McGervey, R.J., Sexstone, A.J., Christ, M.J., Foster, C.J., Adams, M.B. (1998). Nitrous oxide production in two forested watersheds exhibiting symptoms of nitrogen saturation. *Can. J. For. Res.*, 28, 1723–1732.

Richardson, H.L. (1938). The nitrogen cycle in grassland soils: With special reference to the Rothamsted Park Grass Experiment. *J. Agr. Sci.*, 28, 73–121.

Schmitz, A., Sanders, T., Bolte, A., Bussotti, F., Dirnböck, T., Johnson, J., Peñuelas, J., Pollastrini, M., Prescher, A.-K., Sardans, J. et al. (2019). Responses of forest ecosystems in Europe to decreasing nitrogen deposition. *Environ. Pollut.*, 244, 980–994.

Silvertown, J., Poulton, P., Johnson, A.E., Edwards, G., Heard, M., Biss, P.M. (2006). The park grass experiment 1865–2006: Its contribution to ecology. *J. Ecol.*, 94, 801–814.

Smith, H.C. and Miller, G.W. (1987). Managing Appalachian hardwood stands using four regeneration practices, 34- year results. *North. J. Appl. For.*, 4, 180–185.

Stoddard, J.L. (1994). Long-term changes in watershed retention of nitrogen: Its causes and aquatic consequences. In *Environmental Chemistry of Lakes and Reservoirs*, Baker, L.A. (ed.). American Chemical Society, Washington, DC, USA.

Szabó, P. (2015). Historical ecology: Past, present and future. *Biol. Rev.*, 90, 997–1014.

Szabó, P. and Hédl, R. (2011). Advancing the integration of history and ecology for conservation. *Conserv. Biol.*, 25, 680–687.

Walter, C.A., Raiff, D.T., Burnham, M.B., Gilliam, F.S., Adams, M.B., Peterjohn, W.T. (2016). Nitrogen fertilization interacts with light to increase *Rubus* spp. cover in a temperate forest. *Plant Ecol.*, 217, 421–430.

Walter, C.A., Adams, M.B., Gilliam, F.S., Peterjohn, W.T. (2017). Non-random species loss in a forest herbaceous layer following nitrogen addition. *Ecology*, 98, 2322–2332.

4

Gaps and Cracks in Land Cover Mapping for Historical Ecology

Francesca Di PIETRO[1], Roger COLY[1], Clémence CHAUDRON[2], Samuel LETURCQ[1]

[1] *University of Tours, France*

[2] *University of Lorraine, Nancy, France*

4.1. Introduction

Land cover mapping at different historical periods has been a crucial step in historical ecology studies, as it allows the identification of past land cover and the quantification of land cover change, which have an influence on current ecological habitats. Moreover, past land cover, used as a proxy of past land-use, is shown to have an effect on the current composition of plant communities. Indeed, several authors have shown that past rather than present landscape structure can strongly affect present plant community composition (e.g. Lindborg and Eriksson 2004). With regard to land cover history, geographical research on landscape changes, which analyzes changes in land cover over time, has provided a significant contribution. Extensive and accurate land cover mapping, allowing the collection of spatial information on the whole gradient of land cover classes, is particularly valuable. Nevertheless, a large part of historical ecological studies is based on the mapping of one land cover class (often forest patches), mainly based on topographic maps (e.g. Kaim et al. 2016). Despite this limitation, studies on non-forest elements, which require a more detailed knowledge of agricultural land cover, with specific information on both arable land and permanent grassland patches, are growing (e.g. Skaloš and Engstovà 2010). Yet, two types of difficulties for extensive land cover mapping hinder the development of studies in historical ecology: (i) land cover mapping at different historical periods is highly time-consuming and relatively uncertain, as there are few possibilities for checking the accuracy of data sources; (ii) comparison of spatial information from different historical sources, including modern, accurate and detailed databases, leads to a loss of precision, as the less precise source determines the final land cover categorization. These difficulties depend on the type of data available over the period studied. Several authors, mainly geographers and historians, indicate some turning points of land-use change in the modern era, documented in several European countries. Some successive turning points can be defined in rural and urban landscapes evolution (Antrop 2005). (1) In the 19th century, the first part of the century represents the peak in population and therefore in land-use pressure in rural areas (Bergès and Dupouey 2017), as well as the rise of urbanization in the industrial era in urban areas. (2) In the 20th century, the 1950s/1960s represent the beginning of extensive mechanization and intensification of agriculture in rural areas (Meeus et al. 1990), as well as post-war reconstruction in urban areas. The 1970s/1980s represent the moment when the effects of the common agricultural policy (CAP) can be observed in the rural territory, particularly in terms of landscape homogeneity, and the period of suburban extension in cities. In this chapter, we address both main constraints of historical ecology studies: the use of historical maps and the comparison of spatial information from different historical sources.. We address particularly land cover mapping of the 19th century, the 1970s/1980s, and the present period.

4.2. Three main steps of past land cover mapping

Mapping past land cover with geographical information system software generally encompasses three steps: 1) the georeferencing of past maps, which allows the accurate location of images by assigning their geographical coordinates on a georeferenced vector layer; (2) the vectorization of land cover patches and linear elements (e.g. roads and watercourses); and (3) the classification of land cover patches, i.e. the attribution of a specific land cover class to each polygon. The automation of the last two steps is crucial to reduce the time of mapping and could help the study of large areas in a precise way, i.e. by considering the whole range of land cover classes and in particular by distinguishing, in agricultural areas, permanent grasslands from arable lands. The graphic quality of the images influences the probability of successfully automating land cover mapping. Two elements deserve particular attention: (i) the contrast of the colors indicating the heterogeneity of land cover patches and allowing both the vectorization and classification of the land cover patches and (ii) maps should be free of written indications, as these prevent vectorisation. If these two conditions are met, automatic vectorization and classification are possible, by using different methods of image processing and software (e.g. CorelDRAW, ArcScan, an extension of ArcMap or HistMapR package). Otherwise, if the maps' graphic quality does not allow automatic vectorization, manual vectorization is necessary. A slightly less time-consuming and more accurate alternative is to draw the parcel grid manually on tracing paper and then proceed with digitalization, georeferencing and automatic vectorization. The classification, which follows this step in all cases, is then based on visual interpretation of aerial photographs (e.g. Skaloš and Engstovà 2010) or on written information in old cadastral registers (e.g. Skaloš and Kašparová 2012).

4.3. Land cover in the 19th century: the old cadasters

In most European countries, cadasters giving an accurate information on land cover at the parcel scale were produced in the 19th century for fiscal purposes (see Box 4.1). The term "parcel" refers to "any portion of land not divided by material separations, having the same type of culture and belonging to the same owner" (Herbin and Pebereau 1953 in Bartout 2011).

> From the end of the 17th century, a great movement of fiscal reforms began in Europe, in order to resolve the financial deficits linked to the wars in which almost all the states of the continent were engaged. These reforms were directed towards a system of direct taxation: the cadasters. The exceptions were for Switzerland, which had renounced waging external wars at the beginning of the 16th century, and England, which turned towards indirect taxes, due to the opposition of the nobility to direct taxation. During the 18th and 19th centuries, the first cadasters were carried out, based on owners' declarations (as in Luxembourg and Spain), or based on geometrical parcel measurements (as in the Duchy of Milan). The latter, i.e. the parcel cadasters, allowed all parcels to be classified according to the degree of soil fertility. In France, after an unsuccessful attempt in 1763, the parcel cadaster was finally created after the Revolution, by a decision of Emperor Napoleon in 1807. Classifying 100 million parcels, the "Napoleonic cadaster" was completed in more than 40 years (Clergeot 2007).

Box 4.1. *The genesis of national cadastral policies in Europe*

Old cadasters are used in historical ecological studies, particularly in central Europe, northern Europe and in Mediterranean areas. Parcel maps drawn during the previous periods, mainly the 17th and 18th centuries, are sometimes available and used in ecological historical studies, as in Sweden (e.g. Cousins 2001). Before these periods, no systematic parcel maps are available at a national scale (e.g. Szabo and Hédl 2011). This is different for local scales (see Box 4.2).

It was in the 13th century that fiscal documentation was produced to document the land-use in a given territory by describing the parcels that constituted the properties. These were either manorial sources of a private character used to manage the feudal lord's estate (Brunel et al. 2002), or sources of a public character (called "*cadastres*", "*estimes*", "*compoix*", etc.) which served, in Mediterranean Europe, to raise taxes (Abbé et al. 2017). It was not until the end of the Middle Ages that the first parcel plans were produced in support of these tax lists. These plans, which complemented the registers, became increasingly common in the 17th and 18th centuries in Europe.

Box 4.2. *The genesis of land-plot mapping in Europe*

The main benefit of old cadasters is that they include the whole range of land cover classes, and not only a few of them, such as forested, urban and water patches, which are included in topographic maps and often used in historical ecological studies focusing on forest landscapes. Moreover, because of the fiscal reasons for their creation, they are known to be accurate and precise (e.g. Rochel et al. 2017). Nevertheless, their use is highly time-consuming. Two main types of material constitute most of the old cadasters (Figure 4.1).

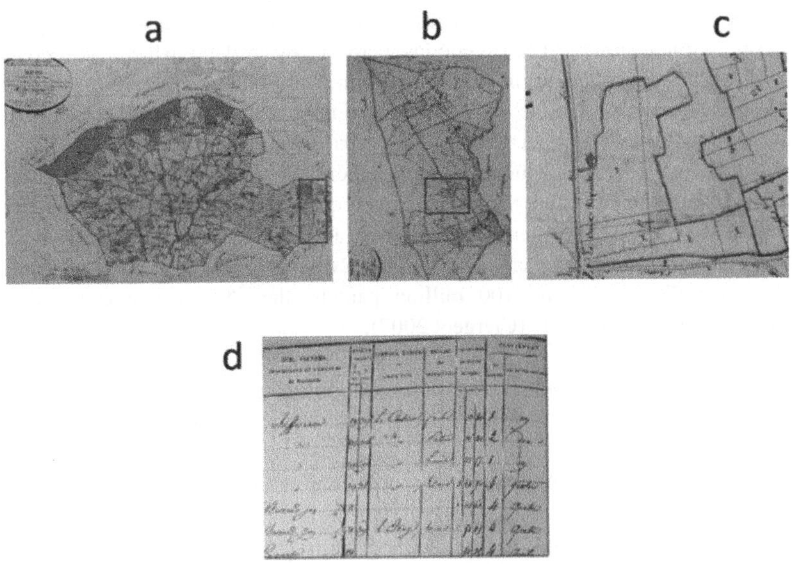

Figure 4.1. *Napoleonic cadaster (1808–1832), municipality of Nouans-les-Fontaines. Joining map (a), parcel map (b), detail of a parcel (c) and parcel register (d) (source: Archives of the Indre-et-Loire department, France). For a color version of this figure, see www.iste.co.uk/decocq/ecology.zip*

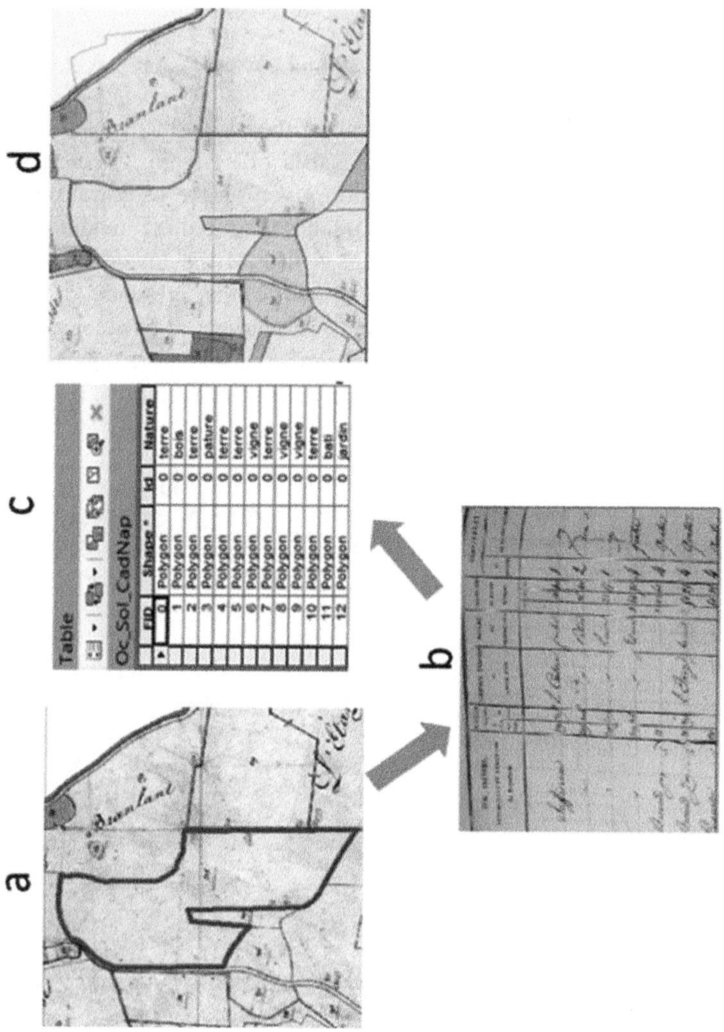

Figure 4.2. *Processing of Napoleonic cadaster (1808–1832). Parcel map (a), compared to parcel register (b), to get land cover specification (c) and mapping (d) (source: Archives of the Indre-et-Loire department, France). For a color version of this figure, see www.iste.co.uk/decocq/ecology.zip*

(i) Parcel maps: in France, parcel maps are drawn up on a scale that varies according to the density of the elements represented (from 1/500 for urban areas to 1/5,000 for rural areas); these digitized parcel maps (raster format) are available on the website of the department archives.

(ii) Parcel registers, which mention the type of land cover of each parcel.

After georeferencing and manual vectorization of the parcel maps, classification of land cover patches is required. This step consists of assigning to each parcel polygon the land cover type from the information given on the cadastral registers. Indeed, the graphic quality of parcel maps does not allow for a direct reading of land cover and automatic mapping of patches. This is due to the low color contrast between land cover types, on the one hand, and to the presence of written information on the maps, with a variety of details on land cover or other elements, on the other hand (Figure 4.2). In order to optimize the cartographic representation, adjacent parcels with the same land cover class can be merged (Figure 4.3).

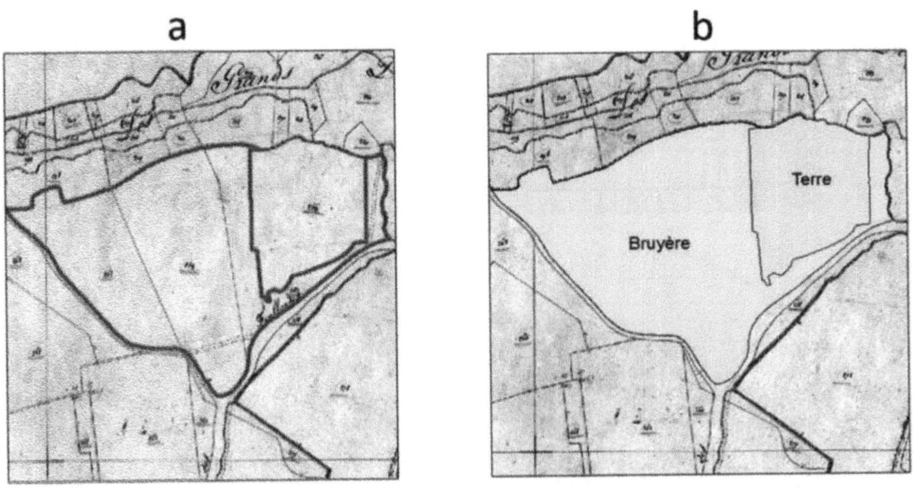

Figure 4.3. *Vectorization of Napoleonic cadaster (1808–1832) by merging parcels with the same land cover type (a, b). "Terre" (Arable) Land, "Bruyère": Heather. For a color version of this figure, see www.iste.co.uk/decocq/ecology.zip*

4.4. Land cover in the 20th century: aerial photographs

The use of vertical aerial photographs to identify land cover in various periods of the 20th century is common in historical ecology and landscape studies (see Gerard et al. 2010). Exhaustive photographic land surveys on a large spatial scale and at regular time intervals were carried out in a majority of European countries and elsewhere, first with black and white photographs (from the 1940s onwards) and then with color photographs (from the 1990s onwards). Moreover, near-infrared photographs, often used in forest-related studies, allow precise identification of chlorophyll activity (color infrared) and wetlands (black and white infrared). The digitized images are often freely available on governmental websites, allowing a wide spatial scope. One of the main interests of aerial photographs is the possibility they offer to identify land cover at several dates, with regular and close dates, and thus capture its changes throughout the 20th century. In our study (Chaudron et al., 2018), the methodology of land cover mapping is based on black and white aerial photographs for the years 1978 and 1980 from National Geographic Institute databases. Each land cover patch is outlined using a semi-automatic vectorization method with CorelDRAW software, and visually checked. Thus, after georeferencing, the semi-automatic vectorization is carried out through two steps (Figure 4.4): (1) an initial vectorization using an image processing software (CorelDRAW v7); (2) then, after a new georeferencing of the vectorized image using the already georeferenced photo, corrections of the vectorized image were necessary to (i) avoid overlaps and gaps, (ii) merge unfairly divided plots or split unfairly aggregated plots, (iii) eliminate polygons that are too small by grouping polygons with the same type of land cover: a threshold of 0.5 ha was chosen to simplify the operation without losing much surface area and (iv) integrate linear elements (road and river networks etc.). The past land cover was then defined based on visual photointerpretation.

We observe that the process of semi-automatic cartography includes, in each of the two main steps, vectorization and photointerpretation, two kinds of difficulties, which limit their use. First, the vectorization produces some extra elements (lines) or deficient elements (gaps) as well as overlaps within and between polygons; moreover, as the distinction between different crops is not always clear on the photo, manual corrections are necessary in order to increase the precision of parcel edges. Second, the photointerpretation of land cover is hampered by the uncertainty of the attribution of a polygon to a land cover class, as the contrast of colors is not always intense and clearly legible. In particular, if the aerial photographs used are not digitized, but are printed on paper, nuances of photographic development and subsequent colors can change from one photo to another: this heterogeneity creates artificial differences and prevents correct vectorization and photointerpretation. Although it is easier to automate, as color contrast between land cover patches is higher than in old cadasters' parcel maps, and there are no written indications on the image, the use of this source of information on land cover seems paradoxically less reliable than the old cadasters, which clearly indicate land cover on parcel registers.

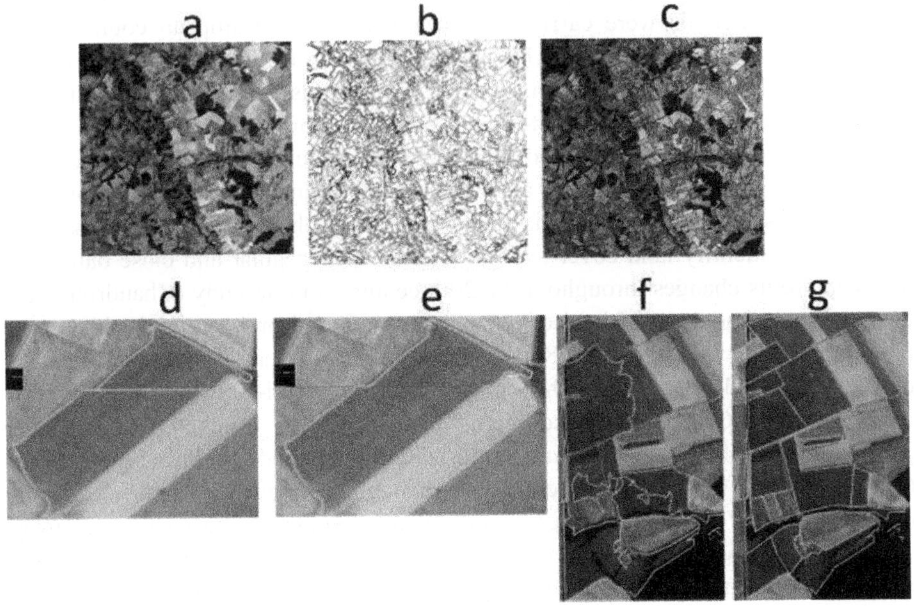

Figure 4.4. *Processing of aerial photographs (date: 1978). Top: vectorization; georeferenced photo (a), first vectorization using an image processing software (b), new georeferencing of the vectorized image using the already georeferenced photo (c). Bottom: corrections of the vectorized image. Unfairly divided plots (d), merging unfairly divided plots (e), unfairly aggregated plots (f) and splitting unfairly aggregated plots (g). For a color version of this figure, see www.iste.co.uk/decocq/ecology.zip*

4.5. Present land cover: modern databases

In combination with past land cover mapping, current land cover mapping is often used in historical ecology studies to measure changes in land cover over time and to compare current and past land cover in terms of their relationship to biodiversity. Contrary to other historical periods, the current period benefits, for land cover mapping, from databases that are already georeferenced, vectorized and informed, which avoid using the three main steps of past land cover mapping mentioned above (georeferencing, vectorization, classification). However, to obtain complete and detailed information, not limited to the main land cover classes such as forested, urban and aquatic areas, it is necessary to integrate into a geographic information system a set of databases from different sources. In particular, information on agricultural land must be added to information on the main land cover classes (forested, urban and aquatic areas) generally available on government websites (in France, this includes the National Geographic Institute's BDTopo®). In France, localized information on agricultural land cover is gathered in the graphic parcels register (GPR, "Registre Parcellaire Graphique"), a database on crops on localized parcels, which is derived from annual declarations made by farmers in respect of the CAP. This source of information, which is made available annually by the Ministry of Agriculture in open access, is valuable for its accuracy on crops, but it requires two types of corrections. On the one hand, information on crop type is available for the main crop of a crop block (set of adjacent parcels used by one farmer) and not at the parcel scale itself: only the main crop per block is shown. It is therefore often necessary to split the crop block into several parcels, by using cadastral digitized maps and by visually checking the orthorectified digital photographs, to obtain information on land cover at the parcel scale. On the other hand, there is missing information about the crop types of some parcels, as some crops are unreported to the CAP (generally permanent grassland or vines). The photointerpretation of the orthophotos is therefore essential to fill these gaps. For both the corrections, it is therefore essential that the GPR and the orthophotos be of the same year. In addition, it should be noted that this crop information, intended for the payment of agricultural subsidies, is generally excessively detailed for historical ecology studies (e.g. the GPR distinguishes winter barley crops from winter wheat crops, etc.). The aggregation of crop classes may be necessary, particularly when unifying the final land cover classification among several periods and sources (see below). Finally, the compatibility between these two main databases (general land cover and specifically agricultural land cover) must be ensured by choosing a compatible date, and by excluding possible overlap between polygons from different databases. Alternatively, maps derived from remote sensing of satellite images are available, such as the maps of the Biosphere's Centre for Spatial Studies, based on Landsat-8 data (Inglada et al. 2017). These raster images can be easily vectorized and their photointerpretation refined by the databases mentioned above.

4.6. From different sources to one land cover typology

From the beginning of the investigations, it is recommended to establish a single land cover classification compatible with the different sources. Indeed, after having mapped the land cover at different dates and from different sources, it is necessary to unify the information on land cover by a single nomenclature. This step, which necessarily leads to a loss of information compared to the most precise information (generally the most recent), is nevertheless essential. It involves compromises about precision, as the less precise source of land cover information leads to the final categorization. It also involves simplification of land cover information, and decrease in the number of unified land cover classes. It should be noted that the parcel registers of the old cadasters include fairly detailed land cover classes (Table 4.1), allowing all parcels to be classified according to the degree of soil fertility for that period, making the establishment of fiscal rules possible. It is worth noting, for example, the plurality of precise terms used to indicate permanent grassland ("Meadow", "Pasture"[1]), semi-natural areas ("Heathland", "Moorland", "Marshes") and uncultivated areas ("Wasteland", "Fallow land" or "terre vaine"). The latter term, "terre vaine", indicates an arable land which is not sown at the time of the cadastral survey and may be widely open to grazing by village herds if the specific range pasturing system ("vaine pâture") is applied. Each land cover class represented at that time a specific level of soil fertility and subsequent income, which is not immediately clear today, especially since some classes have almost disappeared (e.g. the classes "Planted pasture"[2] and "Hemp"). These classifications must be understood according to the uses implemented during the specific historical period studied, in order to integrate them correctly into a diachronic classification, which requires historical information on past agricultural practices.

CODE	Land cover class	CODE	Land cover class	CODE	Land cover class
1	Arable land	8	Hemp	15	Wasteland
2	Meadows	9	Heathland	16	Pond
3	Garden (kitchen garden)	10	Moorland	17	Ditch
4	Vine	11	Marshes (flooded area)	18	Building
5	Orchard	12	Tree nursery	19	Cemetery
6	Planted pasture	13	Forest	20	Quarry
7	Pasture	14	Fallow land (terre vaine)		

Table 4.1. *Land cover classes for the parcels of the study site, Napoleonic cadaster (1808–1832)*

1. The terms "meadow" and "pasture" refer to grazed grassland, but the distinction between the two terms shows a nuance that is difficult to define for the past periods. It is possible that the word "meadow" refers to a sown and possibly mowed pasture, while the word "pasture" refers to a grazed area where spontaneous vegetation grows. See Brumont, F. (2008). *Prés et pâtures en Europe occidentale*. Presses universitaires du Mirail, Toulouse.

2. Planted pasture probably refers to a grassy area planted with fruit trees, used as pasture for livestock.

4.7. Conclusion

The acquisition of localized and accurate information on land cover over time is a crucial step in understanding not only landscape changes, but also changes in ecological patterns, and the legacy that past landscapes represent for present biodiversity. This step is time-consuming, and semi-automatic mapping methods are still in their early stages; they are uncertain and require careful adaptation to the specificities of the information sources. This explains why studies in historical ecology are somewhat limited, being confined to small areas, and restricted to coarse land cover categories (e.g. urban, forested, aquatic, agricultural areas). This restriction does not allow the identification of key categories such as permanent grasslands, indiscriminately associated with arable land in a broad category of "agricultural land". However, semi-automatic vectorization and photointerpretation methods are developing considerably and are leading to the emergence of the field of historical geographical information systems. In the near future, this would undoubtedly help overcome the current hurdles in historical land cover mapping and help to achieve large-scale studies in historical ecology.

4.8. References

Abbé, J.-L., Hautefeuille, F., Jaudon, B., Le Pottier, J., Olivier, S. (2017). Estimes, compoix et cadastres. *Histoire d'un patrimoine commun de l'Europe méridionale*. Le Pas d'Oiseau, Toulouse.

Antrop, M. (2005). Why landscapes of the past are important for the future. *Landscape and Urban Planning*, 70(1–2), 21–34 [Online]. Available at: https://doi.org/10.1016/j.

Bartout, P. (2011). L'apport du cadastre napoléonien aux problématiques spatiales des retenues d'eau. *Revue géographique de l'Est*, 51(3–4). doi: https://doi.org/10.4000/rge.3382.

Bergès, L. and Dupouey, J. (2017). Écologie historique et ancienneté de l'état boisé : concepts, avancées et perspectives de la recherche. *Revue forestière française*, 4–5(1), 297–318.

Brumont F. (2008) *Prés et pâtures en Europe occidentale*. Presses universitaires du Mirail, Toulouse.

Brunel, G., Guyotjeannin, O., Moriceau, J.-M. (2002). Terriers et plans-terriers du XIIIe au XVIIIe siècle. *Actes du Colloque de Paris* (23–25 septembre 1998). Association d'Histoire des sociétés rurales/École nationale des Chartes, Caen-Paris.

Chaudron, C., Perronne, R., Bonthoux, S., Di Pietro, F. (2018). A stronger influence of past than present landscape structure on plant communities of road-field boundary. *Acta Oecologica*, 92(November 2017), 85–94. doi: 10.1016/j.actao.2018.08.009.

Clergeot, P. (2007). *1807 – Un cadastre pour l'Empire. Cent millions de parcelles en France.* Publi-Tope, Paris.

Cousins, S.A.O. (2001). Analysis of land-cover transitions based on 17th and 18th century cadastral maps and aerial photographs. *Landscape Ecology*, 16(1), 41–54. doi: 10.1023/A:1008108704358.

Gerard, F., Petit, S., Smith, G., Thomson, A., Brown, N., Manchester, S., Wadsworth, R., Bugar, G., Halada, L., Bezak, P. et al. (2010). Land cover change in Europe between 1950 and 2000 determined employing aerial photography. *Progress in Physical Geography*, 34(2), 183–205.

Inglada, J., Vincent, A., Arias, M., Tardy, B., Morin, D., Rodes, I. (2017). Operational high resolution land cover map image time series. *Remote Sensing*, 9(95), 1–35. doi: 10.3390/rs9010095.

Kaim, D., Kozak, J., Kolecka, N., Ziółkowska, E., Ostafin, K., Ostapowicz, K., Gimmi, U., Munteanu, C., Radeloff, V.C. (2016). Broad scale forest cover reconstruction from historical topographic maps. *Applied Geography*, 67, 39–48. doi: 10.1016/j.apgeog.2015.12.003.

Lindborg, R. and Eriksson, O. (2004). Historical landscape connectivity affects present plant species diversity. *Ecology*, 85(7), 1840–1845.

Meeus, J.H.A., Wijermans, M.P., Vroom, M.J. (1990). Agricultural landscapes in Europe and their transformation. *Landscape and Urban Planning*, 18(3), 289–352.

Rochel, X., Avon, C., Bergès, L., Chauchard, S., Grel, A. (2017). Quelles sources cartographiques pour la définition des usages anciens du sol en France ? *Revue forestière française*, LXIX(4–5), 353–370 [Online]. Available at: https://doi.org/10.4267/2042/67866.

Skaloš, J. and Engstovà, B. (2010). Methodology for mapping non-forest wood elements using historic cadastral maps and aerial photographs as a basis for management. *Journal of Environmental Management*, 91(4), 831–843. doi: 10.1016/j.jenvman.2009.10.013.

Skaloš, J. and Kasparová, I. (2012). Landscape memory and landscape change in relation to mining. *Ecological Engineering*, 43, 60–69. doi: 10.1016/j.ecoleng.2011.07.001.

Szabo, P. and Hedl, R. (2011). Advancing the integration of history and ecology. *Conservation Biology*, 25(4), 680–687. doi: 10.1111/j.1523-1739.2011.01710.x.

5

The Use of Repeat Photography in African Historical Ecology

Michael Timm HOFFMAN and Rick F. ROHDE
University of Cape Town, South Africa

5.1. Repeat photography as an emerging tool in African historical ecology

One goal of historical ecology is to reconstruct past ecological patterns and dynamics (Beller et al. 2020). Knowing how environments have changed over time in response to key drivers and processes helps us to better understand and manage the present. Describing the trajectory in terms of the nature, extent and rate of past change can also indicate what to expect in the future. Repeat photography is a useful tool in the exploration of this past–present–future environmental change continuum (Hoffman et al. 2019).

Repeat photography involves standing on the same location from where an earlier photograph was made and retaking the image as closely matched to the original photograph as possible (Webb et al. 2010). On-site field surveys and detailed post-production analyses of the matched repeat photographs can provide an accurate account of long-term change at a location (Hammond et al. 2020). Observed variations in species and growth forms (e.g. trees, shrubs and grasses) in different parts of the landscape (e.g. slopes, plains, riverine areas) provide additional information about the pattern of change and the kind of pressures that have been placed on the environment over time. If the trajectories observed in the repeat

Historical Ecology,
coordinated by Guillaume DECOCQ.

photograph pairs are replicated at many different locations, then regional (Jacob et al. 2017) or national (Hoffman et al. 2019) syntheses of long-term environmental change are sometimes possible.

Repeat photography has been used globally for decades to evaluate landscape change (Rogers et al. 1984). Even though one of the earliest applications of repeat photography was undertaken in Africa (Shantz and Turner 1958), it has only been in the last few decades that the technique has been more widely applied on the continent. Nearly three quarters of the 60 main studies which use repeat photography to understand environmental change in Africa have been published since 2010. This chapter documents where these studies have been concentrated on the continent and the main themes that have been addressed using this approach. It also describes the strengths and limitations of the technique and points to some promising future directions.

5.2. Repeat photography and landscape change in Africa

5.2.1. *Early contributions*

Shantz and Turner's (1958) account of vegetation change in Africa is one the earliest uses of repeat photography. It documents long-term changes in vegetation at two time steps (1919 and 1956) along a 4,000 km transect from Cape Town to Nairobi. An analysis of repeat photographs was only used again three decades later by Hoffman and Cowling (1990). Matched photographs of the eastern Karoo region of South Africa were used to assess the hypothesis of desertification that had been proposed earlier by Acocks (1953). Since then, the number of studies which have used repeat photography to understand long-term environmental change on the continent have increased in both number and scope. Isolated studies have been undertaken in several different countries including in Madagascar (Kull 2005), Senegal (Herrmann and Tappan 2013), Kenya (Turner et al. 1998; Western 2010), Tanzania (Rohde and Hilhorst 2001) and Eritrea (Lätt 2004). However, there has been a concentration of repeat photography studies in Ethiopia and southern Africa.

5.2.2. *Ethiopia*

Most of the research in Ethiopia has been undertaken in the semi-arid, northern highlands of Tigray by a group of geographers at the University of Ghent in Belgium (e.g. Nyssen et al. 2008, 2014; De Mûelenaere et al. 2014; Jacob et al. 2015). A major focus has been on assessing the extent of forest cover in early colonial environments and how this has been affected by anthropogenic influences

(Meire et al. 2013). The role of protected areas in preventing deforestation in the Simien Mountains has also been evaluated using historical photographs as has the upward shift in the treeline in areas protected from high anthropogenic pressure (Jacob et al. 2017). Significant effort has been expended on land management interventions in the region and repeat photography has been used to document the impact of these interventions on vegetation cover and soil erosion (Frankl et al. 2012, 2015; Nyssen et al. 2015). Historical photographs have also been used in combination with an analysis of fossil pollen from the endorheic Lake Ashenge in North Ethiopia to understand the cyclical nature of deforestation and forest regrowth over the past 400 years (Lanckriet et al. 2015).

5.2.3. *Southern Africa*

Repeat photography studies in southern Africa have focused on Namibia and South Africa. A major emphasis has been on evaluating the regional climate model forecasts which project widespread aridification for the subcontinent in the future. Historical trajectories of ecological patterns and dynamics have been used to assess the direction and magnitude of change suggested in the model outputs. An analysis of 52 photographs originally taken in 1876 along a 1,000 km transect in Namibia showed how both rainfall and land-use affect vegetation (Rohde and Hoffman 2012). Sites with annual rainfall below 250 mm were relatively stable over time, while woody cover had increased significantly at locations with higher annual rainfall totals. An analysis of 100 repeat photographs from the Namib desert provides evidence of climate change related to regional synoptic changes in savanna rainfall and the fogbelt associated with the coastal Benguela upwelling system (Rohde et al. 2019).

In South Africa, long-term changes in vegetation cover in most of the major biomes have been investigated using repeat photography. In some localities in the Fynbos biome, forests have expanded into areas that were previously dominated by fynbos shrublands largely in response to the suppression and lower frequency of fires relative to the 19th century fire regime (Poulsen and Hoffman 2015). The semi-arid, species-rich Succulent Karoo biome has been relatively stable in terms of vegetation cover on the plains and rocky slopes (Hoffman and Rohde 2007; Figure 5.1), while riverine areas have experienced an increase in the cover of woody plants (Hoffman and Rohde 2011). In the eastern Nama–Karoo biome, there has been an increase in grass cover, especially over the last 50 years, in response to an increase in rainfall and a decrease in livestock (Masubelele et al. 2015; Figure 5.2). The Savanna biome has experienced the greatest change in vegetation cover over the 20th century. In both mesic and semi-arid locations, the cover of woody plants has increased significantly (Hoffman and O'Connor 1999; Russell and Ward 2014; Ward et al. 2014; Figure 5.3). In some cases, the increase in woody plant cover has occurred over just a few decades in natural environments as well as on abandoned croplands. Attribution for these changes has been difficult, and a range of drivers have been invoked to explain the increase in woody plants including changes in herbivory, fire, land-use, rainfall, temperature and CO_2 (O'Connor et al. 2014).

Figure 5.1. *Most matched photographs of the Succulent Karoo biome show stability in vegetation cover over time. Top: Pole Evans c. 1920. Bottom: Rohde and Hoffman, October 27, 2006. Calvinia (No. 246). For a color version of this figure, see www.iste.co.uk/decocq/ecology.zip*

A recent analysis of over 1,300 repeat historical photographs of the biomes of southern Africa provides a summary of the main changes that have taken place in the vegetation of the region over the last 100 years (Hoffman et al. 2019). The long-term trends suggest that the direction of vegetation change is different from that which is expected, based on the outputs from most model forecasts. The trajectories evident in most repeat photograph comparisons reflect environments that have not yet been impacted by climate change but rather landscapes that are recovering from the heavy exploitation of the 19th and early 20th centuries.

Figure 5.2. *At the ecotone between the Nama–Karoo and Grassland biomes, there has been an increase in grass cover. Top: Acocks, July 25, 1946. Bottom: Hoffman and Arena November 2018. Blesbokvlakte 2 (No. 1844). For a color version of this figure, see www.iste.co.uk/decocq/ecology.zip*

Figure 5.3. *The mesic savannas have experienced a significant increase in woody plant cover. Top: Edwards. October 24, 1955. Bottom: Puttick, April 19, 2011. Weenen-Mooi River Road, KwaZulu-Natal (No. 108). For a color version of this figure, see www.iste.co.uk/decocq/ecology.zip*

5.3. Long-term change in plant populations as revealed by repeat photography

Most studies have used an analysis of historical photographs to document landscape level changes in the environment. A few studies, however, have also used closely matched repeat photographs to document changes in plant populations over time. For example, Duncan et al. (2006) counted the number of individuals of the desert tree aloe, *Aloidendron pillansii*, evident in 14 matched photographs from 1950 and 2006, to show a 1.4% annual mortality of the species in the Richtersveld, South Africa. Explanations for the decline include age-related natural causes, climate change, grazing and theft. Age-related annual growth rates of individual tree aloes can also be determined from well-matched repeat photographs and used in population models (Hoffman et al. 2010).

Figure 5.4. *Repeat photographs can be used to assess long-term changes in some species such as the population of Aloidendron dichotomum. There were 66 individuals in the original photograph and only eight in the repeat. Other species in the image include the grass, Stipagrostis ciliaris, and the succulent shrub, Euphorbia gummifera. Top: Marloth c. 1930s. Bottom: Hoffman and Jack, May 25, 2008. Garub, Namibia (No. 472). For a color version of this figure, see www.iste.co.uk/decocq/ecology.zip*

Changes in the status of 1,313 live trees of the iconic conifer, *Widdringtonia cedarbergensis*, in the Cederberg mountains in South Africa were also assessed using 87 repeat photo pairs taken between 1931 and 2013 (White et al. 2016). Counts suggest that 74% of the individuals had died and only 3.4% new individuals had recruited into the population over the period of observation. A generalized linear mixed-effects model (GLMM) suggested that several factors were related to the mortality of this species including greater fire frequency and increased temperatures. Individuals growing at lower elevations and in less rocky habitats were also more susceptible to dying. A third example of how repeat photographs can be used to document long-term changes in plant populations is that of Okubamichael et al. (2016) who studied the fate of 626 individual cycads documented in 107 repeat photographs at 53 locations in South Africa. Their results showed that by 2014, only 16% of the original cycads, evident in the photographs, had survived. They attributed the decline primarily to the theft of individuals for the lucrative cycad trade although the harvesting of bark for medicinal purposes was also important in some areas.

The technique is clearly not suitable for all species. However, for those species with individuals that can be easily counted in photographs (Figure 5.4), the use of repeats can provide useful, quantitative evidence of long-term change in populations. When coupled with a GIS analysis of the location of individuals in the landscape, together with information of key drivers of change such as fire and rainfall, relatively robust predictions about the future trajectory of plant populations can be made (White et al. 2016).

5.4. Strengths and limitations

Repeat photography provides an accurate, practical tool to document the nature, extent and rate of change in key elements of the landscape. When the patterns of change are assessed together with other historical sources such as long-term fire, herbivory and rainfall records, the relative influence of both natural and anthropogenic impacts on the environment can be evaluated (Hoffman et al. 2020). A detailed analysis of historical photographs also fills an important gap in the spatial and temporal scales of observation for global change research (Hammond et al. 2020). Repeat landscape photographs cover a greater area than botanical survey plot data but less area than that usually assessed in aerial photographs or satellite images. While repeat photographs do not have the same level of details about species and populations that are contained in plot data, the quality of this information is greater than that which can be measured from aerial photographs and satellite images (Venter et al. 2020). Perhaps, the most significant contribution that historical photographs make to an understanding of environmental change is that they often extend over 100 years or more (Rohde and Hoffman 2012). This temporal depth is greater than the other approaches mentioned above and helps to set baselines for conservation against which management decisions can be evaluated (Gillson 2015; Beller et al. 2020).

One limitation of repeat photography, and one which it shares with many other historical studies, is the problem of "snapshot" resampling (Stuble et al. 2020). This occurs when only two (or a few) time points are sampled over an extended period. When only the start and end points are sampled, as is usually the case in repeat photography studies, inter-annual variance in the response variable (e.g. plant cover, erosion gulley depth) cannot be determined. Because of this, important signals of long-term change are missed or spurious conclusions about the long-term trends in the response variables are reached (McCain et al. 2016). This can be a significant problem for determining the long-term trend in the populations of short-lived species, but for long-lived species, the same individuals are often found in photographs taken decades apart (Rohde and Hoffman 2012). Knowledge of the longevity of species helps significantly in the interpretation of change observed in repeat photographs as does an understanding of the dynamics and drivers of the vegetation type under investigation (Hoffman et al. 2020).

On their own, a set of matched photographs has only limited value. Any change observed between two photographs requires additional information in order to understand both the context for the pattern and to provide an explanation for the differences. The use of archival sources, traveler's journals and climate and population census data, together with detailed on-site field surveys, often helps with this (Hoffman et al. 2020). A thorough ethnography of relevant land users and institutions is also necessary to appreciate the wider political and economic aspects of environmental change as well as the perspectives of local people (Davis 2009; Von Hellerman 2020). However, attributing change to any one factor is difficult, as it is often several, interacting causes which are thought to be responsible for the differences observed between two photographs (Turner et al. 2003).

5.5. Future directions

The techniques associated with measuring changes in different components of the landscape in matched photo pairs have remained relatively constant over the last 20 years. Most studies use a visual assessment of the relative changes in the pair of photographs to determine long-term trends. Often, the assessment relies on an expert rating system involving multiple observers (Jacob et al. 2017; Hoffman et al. 2019). More objective approaches such as point frequency estimates of plant species cover and number as well as automatic image analysis techniques have also been advanced (Hoffman and Todd 2010). However, these latter techniques have not been adopted widely perhaps because of the greater efficiency and accuracy of visual estimations (Vanha-Majamaa et al. 2000). An objective, widely acceptable, rapid technique to estimate the cover of different components in digital images is still needed.

Repeat photographs on their own are extremely useful for understanding environmental change. However, ground-based images often only cover a limited field of view. When used interactively with other methods such as aerial photographs and remotely sensed images, repeat photographs can provide a reference for metrics of vegetation change derived from these other sources, such as Landsat data (Venter et al. 2020). The integration of historical photographs with other long-term approaches to understand environmental change is only likely to grow. Their value in testing some of the forecasts made by global change models will also be more carefully considered in the future (Hoffman et al. 2020).

Finally, the rapid increase in the use of this technique across the continent suggests that it has been widely adopted as an important tool in global change research. Photographs provide a visually accessible account of long-term change that can be understood and appreciated by a wide range of interested observers. Potential for even greater use has been promoted via the development of citizen science networks (Hammond et al. 2020). The success of rePhotoSA (http://rephotosa.adu.org.za), a citizen science project which encourages members of the public to download and retake historical landscape photographs of southern Africa (Scott et al. 2018), suggests that a wider adoption of the technique is possible. An expanded, easily accessible collection of repeat photographs provides information of long-term environmental change and has obvious implications for ecosystem management.

5.6. References

Acocks, J.P.H. (1953). Veld types of South Africa. *Mem. Bot. Surv. Sth. Afr.*, 28, 1–192.

Beller, E.E., McClenachan, L., Zavaleta, E.S., Larsen, L.G. (2020). Past forward: Recommendations from historical ecology for ecosystem management. *Glob. Ecol. Conserv.*, 21, e00836.

Davis, D.K. (2009). Historical political ecology: On the importance of looking back to move forward. *Geoforum*, 40, 285–286.

De Mûelenaere, S., Frankl, A., Haile, M., Poesen, J., Deckers, J., Munro, N., Veraverbeke, S., Nyssen, J. (2014). Historical landscape photographs for calibration of landsat land use/cover in the northern Ethiopian highlands. *Land Degrad. Dev.*, 25, 319–335.

Duncan, J., Hoffman, M.T., Rohde, R.F., Powell, E., Hendricks, H. (2006). Long-term population changes in the giant quiver tree, Aloe pillansii in the Richtersveld, South Africa. *Plant Ecol.*, 185, 73–84.

Frankl, A., Poesen, J., Deckers, J., Haile, M., Nyssen, J. (2012). Gully head retreat rates in the semi-arid highlands of northern Ethiopia. *Geomorphology*, 173, 185–195.

Frankl, A., Poesen, J., Moeyersons, J., Nyssen, J. (2015). Gully development in the Tigray highlands. In *Landscapes and Landforms of Ethiopia*, Billi, P. (ed.). Springer, Dordrecht.

Gillson, L. (2015). *Biodiversity Conservation and Environmental Change: Using Palaeoecology to Manage Dynamic Landscapes in the Anthropocene*. Oxford University Press, Oxford.

Hammond, W.M., Stone, M.E.B., Stone, P.A. (2020). Picture worth a thousand words: Updating repeat photography for 21st century ecologists. *Ecol. Evol.*, 10(1), 14113–14121.

Herrmann, S. and Tappan, G.G. (2013). Vegetation impoverishment despite greening: A case study from central Senegal. *J. Arid Environ.*, 90, 55–66.

Hoffman, M.T. and Cowling, R.F. (1990). Vegetation change in the semi-arid eastern Karoo over the last 200 years: An expanding karoo-fact or fiction? *Sth. Afr. J. Sci.*, 86, 286–294.

Hoffman, M.T. and O'Connor, T. (1999). Vegetation change over 40 years in the Weenen/Muden area, KwaZulu-Natal: Evidence from photo-panoramas. *Afr. J. Range For. Sci.*, 16, 71–88.

Hoffman, M.T. and Rohde, R.F. (2007). From pastoralism to tourism: The historical impact of changing land use practices in Namaqualand. *J. Arid Environ.*, 70, 641–658.

Hoffman, M.T. and Rohde, R.F. (2011). Rivers through time: Historical changes in the riparian vegetation of the semi-arid, winter rainfall region of South Africa in response to climate and land use. *J. Hist. Biol.*, 44, 59–80.

Hoffman, M.T. and Todd, S.W. (2010). Using fixed-point photography, field surveys and GIS to monitor environmental change in Riemvasmaak, South Africa. In *Repeat Photography: Methods and Applications in the Natural Sciences*, Webb, R., Boyer, D., Turner, R. (eds). Island Press, Washington DC.

Hoffman, M.T., Rohde, R.F., Duncan, J., Kaleme, P. (2010). Repeat photography, climate change and the long-term population dynamics of tree aloes in southern Africa. In *Repeat Photography: Methods and Applications in the Natural Sciences*, Webb, R., Boyer, D., Turner, R. (eds). Island Press, Washington DC.

Hoffman, M.T., Rohde, R.F., Gillson, L. (2019). Rethinking catastrophe? Historical trajectories and modelled future vegetation change in southern Africa. *Anthropocene*, 25, 100189.

Hoffman, M.T., Rohde, R.F., Gillson, L. (2020). Further comments on analysing trajectories of vegetation and landscape change in southern Africa from historical field photographs. *Anthropocene*, 32, 100274.

Jacob, M., Frankl, A., Beeckman, H., Mesfin, G., Hendrickx, M., Guyassa, E., Nyssen, J. (2015). North Ethiopian Afro-Alpine tree line dynamics and forest-cover change since the early 20th century. *Land Degrad. Dev.*, 26, 654–664.

Jacob, M., Frankl, A., Hurni, H., Lanckriet, S., De Ridder, M., Guyassa, E., Beeckman, H., Nyssen, J. (2017). Land cover dynamics in the Simien Mountains (Ethiopia), half a century after establishment of the National Park. *Reg. Environ. Change*, 17(3), 777–787.

Kull, C.A. (2005). Historical landscape repeat photography as a tool for land use change research. *Nor. Geogr. Tidsskr.*, 59, 253–268.

Lanckriet, S., Rucina, S., Frankl, A., Ritler, A., Gelorini, V., Nyssen, J. (2015). Nonlinear vegetation cover changes in the north Ethiopian highlands: Evidence from the Lake Ashenge closed basin. *Sci. Total Environ.*, 536, 996–1006.

Lätt, L. (2004). Eritrea re-photographed. Landscape changes in the Eritrean highlands 1890–2004. Thesis, University of Bern, Switzerland.

Masubelele, M., Hoffman, M.T., Bond, W. (2015). Biome stability and long-term vegetation change in the semi-arid, south-eastern interior of South Africa: A synthesis of repeat photo-monitoring studies. *S. Afr. J. Bot.*, 101, 139–147.

McCain, C.T., Szewczyk, T., Knight, K.B. (2016). Population variability complicates the accurate detection of climate change responses. *Glob. Change Biol.*, 22, 2081–2093.

Meire, E., Frankl, A., De Wulf, A., Haile, M., Deckers, J., Nyssen, J. (2013). Land use and cover dynamics in Africa since the nineteenth century: Warped terrestrial photographs of north Ethiopia. *Reg. Environ. Change*, 13, 717–737.

Nyssen, J., Poesen, J., Haregeweyn, N., Parsons, T. (2008). Environmental change, geomorphic processes and land degradation in tropical highlands. *Catena*, 75, 1–4.

Nyssen, J., Frankl, A., Haile, M., Hurni, H., Descheemaeker, K., Crummey, D., Ritler, A., Portner, B., Nievergelt, B., Moeyersons, J. (2014). Environmental conditions and human drivers for changes to north Ethiopian mountain landscapes over 145 years. *Sci. Total Environ.*, 485, 164–179.

Nyssen, J., Frankl, A., Zenebe, A., Deckers, J., Poesen, J. (2015). Land management in the northern Ethiopian highlands: Local and global perspectives, past, present and future. *Land Degrad. Dev.*, 26, 759–764.

O'Connor, T.G., Puttick, J.R., Hoffman, M.T. (2014). Bush encroachment in southern Africa: Changes and causes. *Afr. J. Range For. Sci.*, 31, 67–88.

Okubamichael, D.Y., Jack, S., Bösenberg, J.D.W., Hoffman, M.T., Donaldson, J.S. (2016). Repeat photography confirms alarming decline in South African cycads. *Biodivers. Conserv.*, 25, 2153–2170.

Poulsen, Z. and Hoffman, M.T. (2015). Changes in the distribution of indigenous forest in Table Mountain National Park during the 20th century. *S. Afr. J. Bot.*, 101, 49–56.

Rogers, G.F., Malde, H.E., Turner, R.M. (1984). *Bibliography of Repeat Photography for Evaluating Landscape Change*. University of Utah Press, Salt Lake City.

Rohde, R.F. and Hilhorst, T. (2001). After the fall: Political ecology and environmental change in the lake manyara basin, Tanzania. Report, IIED Drylands Programme, London.

Rohde, R.F. and Hoffman, M.T. (2012). The historical ecology of Namibian rangelands: Vegetation change since 1876 in response to local and global drivers. *Sci. Total Environ.*, 416, 276–288.

Rohde, R.F., Hoffman, M.T., Durbach, I., Venter, Z., Jack, S. (2019). Vegetation and climate change in the Pro-Namib and Namib desert based on repeat photography: Insights into climate trends. *J. Arid Environ.*, 165, 119–131.

Russell, J. and Ward, D. (2014). Vegetation change in northern KwaZulu-Natal since the Anglo-Zulu war of 1879: Local or global drivers? *Afr. J. Range For. Sci.*, 31, 89–105.

Scott, S.L., Rohde, R.F., Hoffman, M.T. (2018). Repeat landscape photography, historical ecology and the wonder of digital archives in southern Africa. *Afr. Res. Doc.*, 131, 35–47.

Shantz, H.L. and Turner, B.L. (1958). *Photographic Documentation of Vegetational Changes in Africa Over a Third of a Century*. University of Arizona Press, Tucson.

Stuble, K.L., Bewick, S., Fisher, M., Forister, M.L., Harrison, S.P., Shapiro, A.M., Latimer, A.M., Fox, L.R. (2020). The promise and perils of resurveying to understand global change impacts. *Ecol. Monogr.*, Article e01435, 1–14.

Turner, R.M., Ochung, H.A., Turner, J.B. (1998). *Kenya's Changing Landscape*. University of Arizona Press, Tucson.

Turner, R.M., Webb, R.H., Bowers, J.E., Hastings, J.E. (2003). *The Changing Mile Revisited: An Ecological Study of Vegetation Change with Time in the Lower Mile of an Arid and Semiarid Region*. University of Arizona Press, Tucson.

Vanha-Majamaa, I., Salemaa, M., Tuominen, S., Mikkola, K. (2000). Digitized photographs in vegetation analysis – A comparison of cover estimates. *Appl. Veg. Sci.*, 3, 89–94.

Venter, Z., Scott, S.L., Desmet, P.G., Hoffman, M.T. (2020). Application of Landsat-derived vegetation trends over South Africa: Potential for monitoring land degradation and restoration. *Ecol. Indic.*, 113, 106206.

Von Hellerman, P. (2020). Partial stories: Repeat photography, narratives and environmental change in Tanzania. *Vis. Anthropol.*, 33, 363–391.

Ward, D., Hoffman, M.T., Collocott, S.J. (2014). A century of woody plant encroachment in the dry Kimberley savanna of South Africa. *Afr. J. Range For. Sci.*, 31, 107–121.

Webb, R.H., Boyer, D.E., Turner, R.M. (eds) (2010). *Repeat Photography: Methods and Applications in the Natural Sciences*. Island Press, Washington DC.

Western, D. (2010). People, elephants, and habitat: Detecting a century of change using repeat photography. In *Repeat Photography: Methods and Applications in the Natural Sciences*, Webb, R., Boyer, D., Turner, R., (eds). Island Press, Washington DC.

White, J.D.M., Jack, S.L., Hoffman, M.T., Puttick, J., Bonora, D., February, E.C. (2016). Collapse of an iconic conifer: Long-term changes in the demography of *Widdringtonia cedarbergensis* using repeat photography. *BMC Ecol.*, 16, 53.

6

Remote Sensing for Historical Ecology

Pierre-Alexis HERRAULT[1] and David SHEEREN[2]

[1] *University of Strasbourg, France*

[2] *University of Toulouse, France*

6.1. Introduction

Historical ecology involves reconstructing past landscapes and/or their spatio-temporal trajectories to assess the effects of landscape history on current biodiversity patterns (Jackson and Sax 2010). Given the wide range of species response to disturbance, both spatially and temporally, remote sensing offers many advantages for historical ecology through the vast amount of available spatial data and multiple processing approaches.

This chapter provides an inventory of the available remote sensing approaches for historical ecology at various spatio-temporal scales in relation to the existing and accessible spatial data. This review is not intended to be exhaustive but offers readers a glimpse of the possibilities and the stage of development of these approaches. Two aspects will be highlighted throughout this chapter: (1) the extraction of past landscapes from various available spatial sources and (2) the production of spatio-temporal variables to model the effects of landscape history on current biodiversity.

Historical Ecology,

6.2. Landscape spatio-temporal changes as a proxy of biodiversity

Present-day biodiversity often reflects past land-use or past climate because of a possible delay in the response of certain species to habitat disturbance (Kuussaari et al. 2009). This interval depends on the changes themselves (i.e. intensity and cyclicity) as well as species traits, namely their dispersal abilities. To assess the time lag between changes in a landscape and changes in the associated populations, studies usually rely on past and current spatial data that help highlight spatio-temporal trajectories over large spatial extents and long periods.

In landscape ecology, biodiversity is most often studied from the perspective of ecological habitats. Several fundamental ecological theories (e.g. island biogeography, species-energy, species-heterogeneity) describe the relationships between habitats and biodiversity, and these theories serve as a framework for historical ecology. Therefore, a crucial step consists of describing specific habitat properties by using adapted metrics. These properties can be distinguished as follows:

– structural (or morphological) properties, which characterize the composition (nature) and configuration (morphology, spatial organization) of a habitat;

– phyletic (or inherited) properties, which characterize the links between patches of homologous habitats at different time slices;

– functional (or spectral) properties, which characterize the behavior of habitats (productivity, phenology, etc.) that can be derived from high temporal resolution signals that are recorded by satellite- and aircraft-based sensors.

The first two categories are available through the mapping of past landscapes at different times (landscape states). Identifying the structure of a habitat requires landscape metrics that permit characterizing habitat morphology (patch scale) or spatial organization (patch-neighborhood or landscape scale). The phyletic characteristics of a habitat and its evolution are obtained by linking homologous patches across different time slices. Combining structural and phyletic properties will ultimately provide evidence of past processes, such as the fragmentation or simplification of landscapes. Finally, functional properties may be analyzed through intra-annual or interannual biophysical variables that are calculated at a high temporal frequency, as these variables are proxies for monitoring habitats and habitat function over a relatively short period (seasonal cycles). Consequently, high temporal resolution data are often required for measuring functional properties and are, by default, available for more recent periods.

6.3. Mapping landscapes at different dates

Prior to describing the characteristics of ecological habitats, habitat patches must first be extracted from the available spatial data for various time slices (Figure 6.1). This initial step can be tedious depending on the quality (geometric, semantic, temporal, spectral) of the available sources, thereby limiting the reliability of results over extended areas. Here, we present various existing automated methods that facilitate the extraction of landscape patches from commonly used spatial data or that are potentially useful in historical ecology.

6.3.1. *Airborne laser scanning data*

Given the ability of LiIDAR to penetrate the forest canopy and detect micro-features, LiDAR-derived data (Digital Terrain Model, DTM) can be used to identify past ecosystems or provide additional information for previously discovered sites (Dupouey et al. 2002). Because of the limited properties of these data, for example, gray levels and variations in lighting, this step is usually undertaken manually for historical ecology studies. This limitation is also related to a lack of collaboration between historical ecology and archaeology, a field in which LiDAR data have been exploited by researchers for many years.

Several approaches for detecting relatively old ecological habitats have been proposed (Lambers et al. 2019), such as template-matching or knowledge-based algorithms as well as object-based approaches or machine-learning methods. Of these approaches, machine learning provides the best performance and consists of calibrating the detection models using a set of learning samples having features calculated from the digital terrain model. The use of decision trees, for example, random forest, is generally successful; however, the results can be limited when transposing the model outputs to another shape or to a different region. Current deep learning methods show promise for the automatic mapping of archaeological relics identified from LiDAR-based digital terrain models (Trier *et al.* 2019). Nonetheless, the calibration of these neural networks requires an extensive set of learning examples, an aspect that is not evident owing to the rare nature of archaeological objects. The available image datasets (e.g. AlexNet, ImageNet) from which the networks can be pretrained also contain insufficient examples for reproducing the full variability of the forms assumed by these archaeological objects.

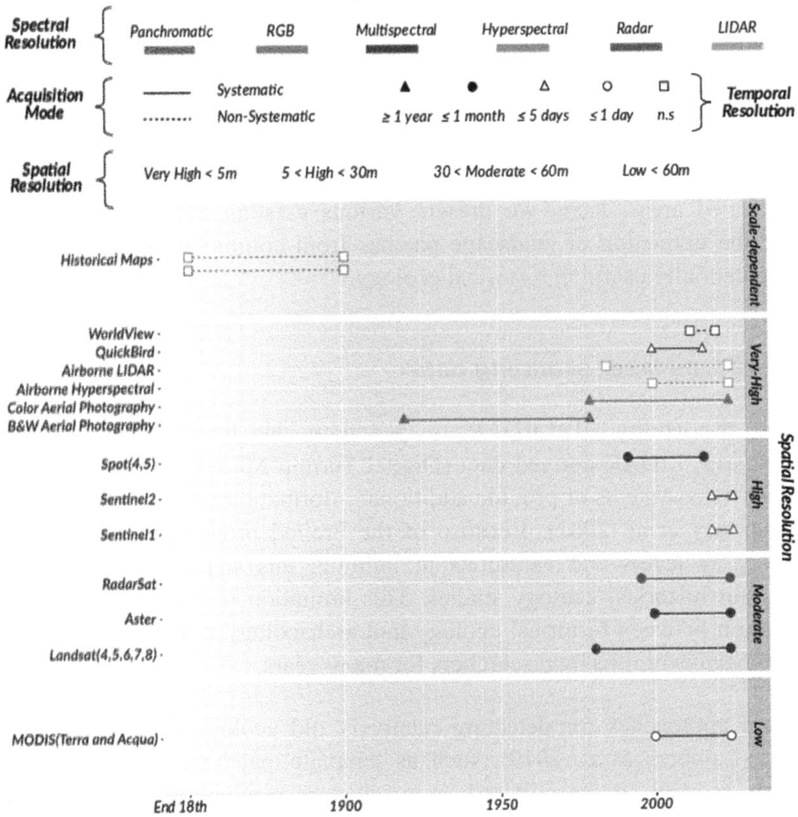

Figure 6.1. *Non-exhaustive overview of some of the major spatial sources available for landscape/historical ecology studies at the landscape scale. RGB indicates the red-green-blue spectral domain. n.s. indicates a "non-significant" temporal resolution because of the non-systematic acquisition mode of the target mission. For a color version of this figure, see www.iste.co.uk/decocq/ecology.zip*

6.3.2. *Historical maps*

Old maps are regularly used in historical ecology (Cousins et al. 2015). They provide a snapshot of landscapes from the 17th to 19th centuries and are commonly the earliest time slice that is integrated into spatio-temporal landscape datasets. The traditional approach of capturing the cartographic objects found on old maps relies on user intervention (for digitizing). For numerous years, automated data-capture techniques were developed to try and establish reproducible procedures. Most of the developed approaches, however, were specific to only one particular map and were generally not applicable to other historical maps.

Recent work appears to be converging towards a standardization of the processing chains (Herrault et al. 2013; Auffret et al. 2017). First, an initial cleaning step is required systematically to eliminate elements that overlap the sought-after data on the maps, for example, hatching, topographic contours, text. Smoothing algorithms based on convolution or mathematical morphology are generally used. Second, a classification step (object or pixel) is applied to assign each pixel or object to a land-use class using supervised algorithms. More recently, neural networks (deep learning) based on convolution kernels have been proposed for this step (Chen et al. 2021). The obtained results show promise and a higher quality than those obtained through traditional machine learning. Nonetheless, the parameterization of these networks and the amount of required training data remain obstacles to their wider application in historical ecology. Finally, a post-classification processing step may be required, depending on the noise observed in the classification outputs. The methods used for this processing are similar to those applied during the initial cleaning step.

6.3.3. *Old aerial photographs*

Historical orthophotographs are primarily used to monitor landscape data from the early 20th century (since the early 1930s) to the first systematic satellite acquisitions in the early 1980s. These photographs provide information about landscapes at a high spatial resolution (approximately 1 m) (Morgan et al. 2017) and have a temporal resolution (time step between aerial surveys over the same geographical area) of about 10 years. Despite these advantages, the delimitation of habitats remains a tedious step because: (1) there is a high degree of heterogeneity between surveys and even between images related to varying specifications and quality (noise problem and vignetting); and (2) these data lack usable attributes, particularly for panchromatic images, making automatic classification a very complex task.

Approaches have been proposed recently to overcome these obstacles. First, a spectralization/colorization process can be applied to enrich spectral attributes of the images, particularly for B\&W photographs. The most effective approach at present is based on deep-learning models, which have considerably improved the colorization results (Goodfellow et al. 2014). The process involves training a neural network using a dataset of paired panchromatic-color images to derive semantics capable of generalizing the colorization to the whole target scene. In particular, generative antagonistic networks (GAN) provide a much higher colorization quality than that obtained with convolutional neural networks (CNN). Significant improvements have also been demonstrated in the classification results obtained from these newly colored images (Poterek et al. 2020). The production of old digital surface models using photogrammetry techniques is also a promising avenue for enriching attributes of old aerial missions and thus facilitate the extraction of elements from these sources (ANR HIATUS, https://anr.fr/Projet-ANR-18-CE23-0025).

Second, object-oriented approaches are most often favored for classifying aerial photographs because they can process very high spatial resolution data. Several studies have also demonstrated the utility of deep learning for the automatic classification of aerial photographs (Sameen et al. 2018). Most rely on pretrained CNN using available image datasets, such as AlexNet or ImageNet. These networks have the advantage of teaching themselves the necessary features to classify at different scales. They are therefore very effective for deriving contextual metrics and often produce superior results to those obtained through traditional methods based notably on textural attributes.

6.3.4. *Satellite images*

An exhaustive review of image satellite classification method is beyond the scope of this chapter. Various recent publications provide an overview of existing approaches (Dhingra and Kumar 2019). From a landscape ecology perspective, two key aspects should be underlined:

– Classification outputs are strongly dependent on spatial, spectral and temporal characteristics of the available satellite missions. These characteristics determine the geometric and semantic quality of the produced land-cover maps.

– Machine-learning methods are currently the standard for image classification. The quality of the learning examples and the selection of features are two critical elements to their success. In recent years, the performance of classification algorithms has rapidly improved through the use of CNN. The quality of the output, however, remains highly dependent on the massive image datasets required for their calibration. Although these datasets are well developed for the RGB domain, they remain relatively poor in examples for the more resolved spectral domains, for example, hyperspectral and radar data.

6.4. Modeling the effects of spatio-temporal changes on present-day biodiversity

The ecological processes structuring biodiversity patterns can be disturbed by forcing events of varying magnitude and intensity (Jackson and Sax 2010). These changes are of several orders: (1) structural changes describing modifications in landscape configuration and composition and (2) functional changes revealing fluctuations in habitat properties, for example, productivity and phenology (Figure 1.2).

6.4.1. *Structural spatio-temporal metrics*

Here, we present the existing structural spatio-temporal metrics according to two major disturbances: landscape fragmentation and landscape homogenization.

6.4.1.1. *Landscape fragmentation*

The fragmentation of landscapes reduces habitat size and heightens habitat isolation (Fahrig 2003). Assessing the effects on biodiversity follows two central ecological theories: the theory of island biogeography, including species–area relationships, and metapopulation theory.

The morphological description of habitat patches makes it possible to use species–area relationships to assess biodiversity loss or gain in accordance with patch attrition or expansion. This theory is also closely related to island biogeography, which stipulates that at constant colonization rates, the extinction rate of species is likely to decrease (increase) as the area increases (decreases) through a positive effect of size on richness. Larger patches are generally more heterogeneous than smaller ones and therefore favor a greater diversity of resources for different species. Thus, the morphology of habitat patches can be characterized simply using surface, perimeter or shape metrics. This implies the availability of geometric objects (most often polygonal) at different times for one or more habitat types. At the landscape scale, the colonization/extinction dynamics of species also depend on the connectivity of the habitat patches. For a given surface area, the colonization rate of a habitat patch is lower when the patch is further away from the source; the diversity will therefore be higher for a less isolated habitat. The degree of complexity of the connectivity metrics is variable (Kindlmann and Burel 2008), and their use most often depends on the specific research objectives and the semantic quality of land-cover maps, i.e. the information available through the landscape matrix.

The effects of habitat disturbances on biodiversity are most often modeled by comparing the characteristics of past and present landscapes. The difference between current richness and the value predicted by the statistical model on the basis of past landscapes provides an estimate of the post-disturbance species debt or credit. This type of modeling has been widely adopted for different taxa (Figueiredo et al. 2019) but may be questioned because of the possibilities of over- or underestimating extinction processes. Other approaches aim to integrate the temporal trajectory of habitats into species-habitat models by considering habitat changes as a cumulative process. Consequently, in addition to the present habitat characteristics, accounting for variables such as the rate of area change and connectivity may considerably improve the explanation of species richness and abundance (Herrault et al. 2016). Finally, the life history of habitat patches (their phyletic properties; Figure 1.2) over time appears to be a relevant variable for explaining current biodiversity patterns (Ewers et al. 2013). The historical separation of habitat fragments is represented by a landscape "terrageny" in a similar manner as phylogeny represents the evolutionary divergence among species.

6.4.1.2. *Landscape homogenization*

The homogenization of a landscape reduces its capacity to support a large number of species owing to the lack of complementary resources to meet the needs of a large pool of species. This results in a decline in overall diversity (taxonomic diversity) as well as functional diversity, with specialists being replaced gradually by generalist species (Gámez-Virués et al. 2015).

Landscape homogenization is characterized both as a decrease in the diversity of land-use (heterogeneity of composition) and as an increase of habitat patch size (heterogeneity of configuration; Fahrig et al. 2011) The configurational heterogeneity is primarily based on morphological metrics (area, length) and summarized (summed or averaged) at the scale of the neighborhood of the habitat patches. The compositional heterogeneity is measured using indicators of land-use diversity, such as the relative share of land-use in the neighborhood of the patch and/or Shannon metrics, which have the benefit of being less correlated with configuration metrics. Modeling the effects of homogenization on biodiversity is most often undertaken via statistical models relating the characteristics of the current landscape to current biodiversity. Other researchers introduce temporal composition metrics (absolute or relative, for example, ΔShannon, ΔWood) directly into ecological distribution models (Bonthoux et al. 2013). It should be noted that this latter approach remains marginal among the cited publications, notably because of the difficulties in reconstructing past land-use trajectories. Moreover, these metrics imply having a detailed and uniform land-use nomenclature for the considered time trajectory (at different times). This can be a limiting factor for older periods where the available sources are not always of sufficient semantic quality relative to the established nomenclature criteria. Image classification errors can also limit the reliability of derived spatio-temporal trajectories by leading to a loss of information at each time slice. Finally, the strict delimitation of objects in the landscape does not always reveal the details and continuity of a landscape.

Given these latter issues, a continuous approach (including no classification step) has been proposed and contrasts with the above-mentioned discrete vision. These works are based on the spectral variation hypothesis (SVH), which assumes that the spectral dispersion in the image (NDVI, texture or reflectance values) is correlated with the heterogeneity of the habitat patches in the landscape. Many approaches have been proposed to quantify spectral heterogeneity and relate it to alpha and beta diversities (Rocchini et al. 2010). They include (1) dispersion-based metrics, which rely on calculations of the standard deviation or the coefficient of variation of pixel values; (2) spectral distance-based metrics, which involve an initial classification of pixels with non-supervised methods, such as K-means methods, to describe similarities between spectral classes. Landscape heterogeneity is then calculated with diversity indices (e.g. Shannon, Rao; Rocchini et al. 2021). Results suggest that spectroheterogeneity is an effective and quickly produced proxy of biodiversity; however, the quality of its output is strongly dependent on (1) the characteristics of satellite or airborne sensors; (2) scale-matching problems between the remotely sensed and field-based diversity data, and (3) the types of species diversity measures.

6.4.2. *Functional spatio-temporal metrics*

Functional metrics require high temporal resolution spatial data to capture the behavior of ecological habitats (Figure 6.2). These spectral–temporal metrics refer to different aspects of flora and fauna habitat such as productivity, phenology, moisture or snow cover.

Productivity metrics are most often obtained through calculating biophysical variables from various satellite images. Nonetheless, few satellite missions provide sufficient temporal characteristics, for example, frequency and age, to highlight the seasonal behavior of landscapes and their past interannual variations. The MODIS mission, for example, provides high-frequency NDVI data capturing the vegetation photosynthetic activity considered as a proxy of primary productivity. Composite products have therefore been proposed, for example, dynamic habitat indices (DHI), and make it possible to simultaneously assess the role of different factors (vegetation stress (DHI_min), seasonal variations (DHI_var) and total productivity (DHI_cum)) on bird, mammal and amphibian diversity patterns (Radeloff et al. 2019). Few studies, however, have tested DHI interannual trends as a potential explanatory variable of current biodiversity patterns. One reason could be the low spatial resolution of MODIS that may hinder its use owing to a spatial mismatch with field data. Despite a lower temporal resolution (16 days), Landsat data are also relevant for deriving functional productivity metrics. The succession of Landsat sensors (since the early 1980s) and the recent Sentinel 2 missions make a major asset for reconstructing the past annual productivity of habitat patches over a longer period of time.

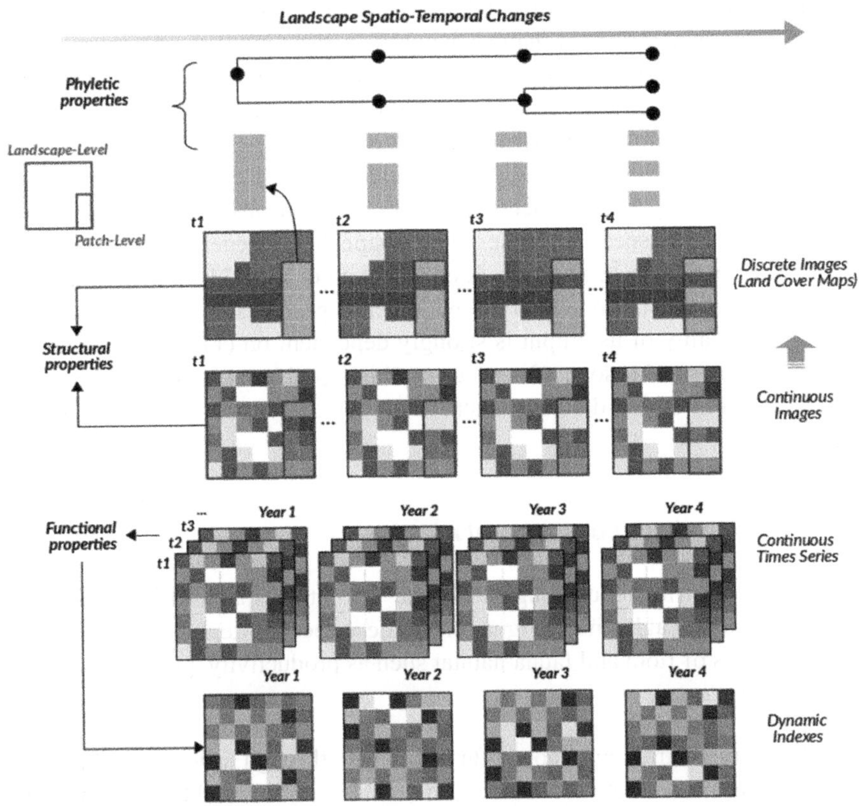

Figure 6.2. *Landscape spatio-temporal reconstruction based on the properties of ecological habitats. For a color version of this figure, see www.iste.co.uk/decocq/ecology.zip*

Spatial and temporal variations in phenology are critical factors in resource availability and habitat use for a wide range of species. Changes in phenological timing can lead to dysfunctions in the interactions between species and, more specifically, to cascading asynchronies across trophic levels (Mayor et al. 2017). Such indices are based on the initial calculation of seasonal indices, such as the day of the start/end of season, peak greening. They can be derived from the intra-annual curves of vegetation indices, for example, EVI and NDVI. Consequently, high temporal resolution satellite data are required to capture these seasonal metrics. For instance, the MODIS mission provides a phenological product (MCD12Q2) having different bands that display these transition dates. The Landsat and Sentinel 2 missions are also well adapted to derive these metrics. Nonetheless, the moderate temporal resolution and low temporal depth remain two limiting factors for reconstructing spatio-temporal phenological trajectories. Developing synergies between these two missions will be critical for future studies in historical ecology.

6.5. References

Auffret, A.G., Kimberley, A., Plue, J., Skånes, H., Jakobsson, S., Waldén, E.,Wennbom, M., Wood, H., Bullock, J.M., Cousins, S.A. et al. (2017). Histmapr: Rapid digitization of historical land-use maps in r. *Meth. Eco. Evo.*, 8(11), 1453–1457.

Bonthoux, S., Barnagaud, J.-Y., Goulard, M., Balent, G. (2013). Contrasting spatial and temporal responses of bird communities to landscape changes. *Oecologia*, 172(2), 563–574.

Chen, Y., Carlinet, E., Chazalon, J., Mallet, C., Duménieu, B., Perret, J. (2021). Combining deep learning and mathematical morphology for historical map segmentation, arXiv preprint arXiv:2101.02144.

Cousins, S.A., Auffret, A.G., Lindgren, J., Tränk, L. (2015). Regional-scale land-cover change during the 20th century and its consequences for biodiversity. *AMBIO A Journal of the Human Environment*, 44(1), 17–27.

Dhingra, S. and Kumar, D. (2019). A review of remotely sensed satellite image classification. *Int. Jour. Elec. Comp. Eng.* (2088–8708), 9(3).

Dupouey, J.-L., Dambrine, E., Laffite, J.-D., Moares, C. (2002). Irreversible impact of past land use on forest soils and biodiversity. *Ecology*, 83(11), 2978–2984.

Ewers, R.M., Didham, R.K., Pearse, W.D., Lefebvre, V., Rosa, I.M., Carreiras, J.M., Lucas, R.M., Reuman, D.C. (2013). Using landscape history to predict biodiversity patterns in fragmented landscapes. *Eco. Lett.*, 16(10), 1221–1233.

Fahrig, L. (2003). Effects of habitat fragmentation on biodiversity. *Ann. Rev. Eco. Evo. Sys.*, 34(1), 487–515.

Fahrig, L., Baudry, J., Brotons, L., Burel, F.G., Crist, T.O., Fuller, R.J., Sirami, C., Siriwardena, G.M., Martin, J.-L. (2011). Functional landscape heterogeneity and animal biodiversity in agricultural landscapes. *Eco. Lett.*, 14(2), 101–112.

Figueiredo, L., Krauss, J., Steffan-Dewenter, I., Sarmento Cabral, J. (2019). Understanding extinction debts: Spatio-temporal scales, mechanisms and a roadmap for future research. *Ecography*, 42(12), 1973–1990.

Gámez-Virués, S., Perovic, D.J., Gossner, M.M., Börschig, C., Blüthgen, N., De Jong, H., Simons, N.K., Klein, A.-M., Krauss, J., Maier, G. et al. (2015). Landscape simplification filters species traits and drives biotic homogenization. *Nat. Com.*, 6(1), 1–8.

Goodfellow, I., Pouget-Abadie, J., Mirza, M., Xu, B., Warde-Farley, D., Ozair, S., Courville, A., Bengio, Y. (2014), Generative adversarial networks. *Advances in Neural Information Processing Systems*. ACM, New York, 26722680.

Herrault, P.-A., Sheeren, D., Fauvel, M., Paegelow, M. (2013). Automatic extraction of forests from historical maps based on unsupervised classification in the CIELab color space. In *Geographic Information Science at the Heart of Europe*, Vandenbroucke, D., Bucher, B., Crompvoets, J. (eds). Springer, Wiesbaden.

Herrault, P.-A., Larrieu, L., Cordier, S., Gimmi, U., Lachat, T., Ouin, A., Sarthou, J.-P., Sheeren, D. (2016). Combined effects of area, connectivity, history and structural heterogeneity of woodlands on the species richness of hoverflies (diptera:Syrphidae). *Land. Eco.*, 31(4), 877–893.

Jackson, S.T. and Sax, D.F. (2010). Balancing biodiversity in a changing environment: Extinction debt, immigration credit and species turnover. *Tren. Eco. Evo.*, 25(3), 153–160.

Kindlmann, P. and Burel, F. (2008). Connectivity measures: A review. *Land. Eco.*, 23(8). 879–890.

Kuussaari, M., Bommarco, R., Heikkinen, R.K., Helm, A., Krauss, J., Lindborg, R., Öckinger, E., Pärtel, M., Pino, J., Rodà, F. et al. (2009). Extinction debt: A challenge for biodiversity conservation. *Tren. Eco. Evo.*, 24(10), 564–571.

Lambers, K., Verschoof-van der Vaart, W.B., Bourgeois, Q.P. (2019). Integrating remote sensing, machine learning, and citizen science in Dutch archaeological prospection. *Rem. Sen.*, 11(7). 794.

Mayor, S.J., Guralnick, R.P., Tingley, M.W., Otegui, J., Withey, J.C., Elmendorf, S.C., Andrew, M.E., Leyk, S., Pearse, I.S., Schneider, D.C. (2017). Increasing phenological asynchrony between spring green-up and arrival of migratory birds. *Sci. Rep.*, 7(1), 1–10.

Morgan, J.L., Gergel, S.E., Ankerson, C., Tomscha, S.A., Sutherland, I.J. (2017). Historical aerial photography for landscape analysis. In *Learning Landscape Ecology*, Gergel, S.E. and Tuner, M.G. (eds). Springer, Cham.

Poterek, Q., Herrault, P.-A., Skupinski, G., Sheeren, D. (2020). Deep learning for automatic colorization of legacy grayscale aerial photographs. *IEEE Jour. Sel. Top. App. Ear. Obs. Rem. Sen.*, 13, 2899–2915.

Radeloff, V., Dubinin, M., Coops, N., Allen, A., Brooks, T., Clayton, M., Costa, G.,Graham, C., Helmers, D., Ives, A. et al. (2019). The dynamic habitat indices (DHIs) from modis and global biodiversity. *Rem. Sen. Env.*, 222, 204–214.

Rocchini, D., Balkenhol, N., Carter, G.A., Foody, G.M., Gillespie, T.W., He, K.S., Kark, S., Levin, N., Lucas, K., Luoto, M. et al. (2010). Remotely sensed spectral heterogeneity as a proxy of species diversity: Recent advances and open challenges. *Eco. Inf.*, 5(5), 318–329.

Rocchini, D., Salvatori, N., Beierkuhnlein, C., Chiarucci, A., de Boissieu, F., Förster, M., Garzon-Lopez, C.X., Gillespie, T.W., Hauffe, H.C., He, K.S. et al. (2021). From local spectral species to global spectral communities: A benchmark for ecosystem diversity estimate by remote sensing. *Eco. Inf.*, 61, 101195.

Sameen, M.I., Pradhan, B., Aziz, O.S. (2018). Classification of very high resolution aerial photos using spectral-spatial convolutional neural networks. *Jour. Sens.*, 2018, Article ID 7195432, 1–12.

Trier, Ø.D., Cowley, D.C., Waldeland, A.U. (2019). Using deep neural networks on airborne laser scanning data: Results from a case study of semi-automatic mapping of archaeological topography on arran, Scotland. *Arch. Pros.*, 26(2), 165–175.

7

Soil Archives: Where Soilscape History Meets Present-day Ecosystems

Boris BRASSEUR[1], Damien ERTLEN[2] and Vincent ROBIN[3]
[1] Jules Verne University of Picardie, Amiens, France
[2] University of Strasbourg, France
[3] University of Lorraine, Metz, France

7.1. Introduction

The interpretation of soil archives, although complex, is nevertheless essential for understanding the impact of soils on present-day environments and deciphering the environmental history of a site. Soil archives are local, although within a continuous soilscape, they allow the study of spatial structures. Compared with sedimentary archives, they do not strictly follow stratigraphic principles. Combining superficial microtopography, buried biological remains, bioturbation and chemical evolution, these archives appear as mixed pieces of old and new disassembled jigsaw puzzles. Such puzzles are snapshots of past environments, and understanding the biological and physicochemical pedogenic processes is the main way to disentangle them. This chapter presents some key elements for understanding the archiving process through to the reconstruction of a fragment of paleoenvironmental history. The presented key elements are associated with our respective experiences, and we do not try to present an exhaustive list.

Historical Ecology,
coordinated by Guillaume DECOCQ.

7.2. Mechanisms of soil archiving and the associated dynamics

7.2.1. *Pedoturbations of biological and physical origins*

When dealing with soil archives, a key issue to consider is that soils are dynamic entities. Biological and/or physical processes induce the mixing of the soil matrix down to a fine spatial scale (i.e. micromorphological scale). Indeed, when environmental conditions permit biological activities in soil, the related bioturbation causes a vertical transfer of elements. These transfers are mainly related to the presence of micropassages in which soil moves downward when they collapse. These spaces result from the activity of small mammals (e.g. *Talpidae*) and earthworms (e.g. *Lumbricidae*). The latter of these also move worm casts – matter excreted by the worms – upward to the soil's surface (Lavelle et al. 2016). The mixing of soil also occurs through physical processes. These processes include tree uprooting, in which soil embedded in the root plate (Šamonil et al. 2010) moves upward through the overturning of the tree trunk and roots. Cryoturbation stems from the freeze–thaw-induced movement of the entire active layer portion of the soil (Stinchcomb et al. 2014). Argillipedoturbation (swelling clay pedoturbation) and seismiturbation (earthquake pedoturbation) are also important physical soil disturbances. Thus, the soil matrix or soil aggregates may move in any direction, depending on the type and intensity of soil biological activities, the on-site vegetation cover, local climate, seasonality, soil clay content and local tectonic activity. These processes sometimes preserve the soil's environmental memory, but they also represent a substantial challenge for researchers trying to read the soil archive.

7.2.2. *Eluviation–Illuviation*

The percolation of water through the soil is an important process in humid climates. This downward movement of water transports fine particles (e.g. clays, fine silt, organic matter) and molecules (e.g. organic matter, oxyhydroxides, carbonates) capable of passing through the pore spaces of the soil matrix. These two components are transported, respectively, in suspension or dissolved in the water as they pass through the pore space. A soil layer subjected mainly to the departure of particles is termed "eluvial" (E horizons) and often undergoes partial discoloration (Figure 7.1). With the downward change in the texture of the horizons (i.e. clay accumulation), the disappearance of larger pores (thus a reduced movement through the soil), the pH gradient of the soil and other physical obstacles (i.e. blocks of limestone), the eluviated soil particles are likely to flocculate and be adsorbed at depth, and the molecules will also precipitate at a deeper soil horizon. The soil area receiving the material transported by gravity water is termed "illuvial" (B horizons).

These eluviation–illuviation dynamics resulting from precipitation will, over the centuries, determine the structure and texture of soils. With these transfers/translocations of particles and molecules, many parameters fundamental to plant growth (water storage capacity of soil horizons, volume density, CEC, resistance to penetration) are influenced over these longer time scales.

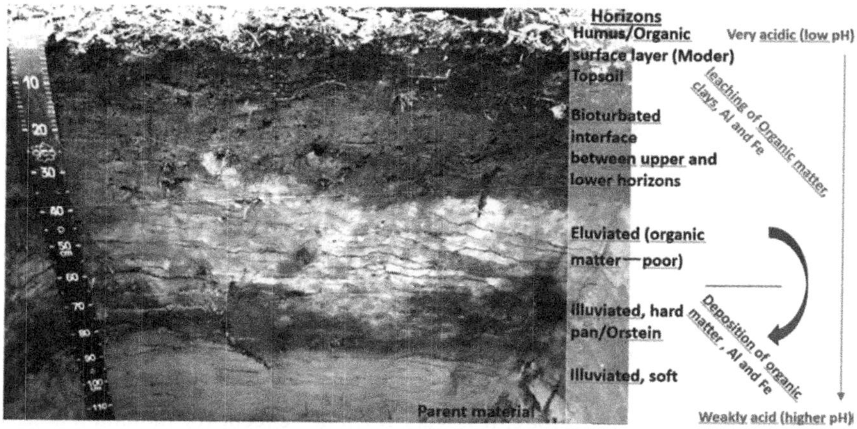

Figure 7.1. *Soil profile showing distinct horizons (labeled on the right-hand side) that mark the eluviation and illuviation of organic matter (note the dark horizontal stripes at approximately 50 cm depth) and metal oxides (approximately 65–85 cm depth) at the expense of the initial parent material. This type of soil, a podzol, is dominated by these vertical redistributions and little bioturbation. For a color version of this figure, see www.iste.co.uk/decocq/ecology.zip*

Human influence through land-use (field, pasture, forest) can be very significant and long-lasting. For example, human modification of pH can affect eluviation–illuviation patterns. In the case of eluviation–illuviation of clays, Ca^{2+} and Al^{3+} ions are flocculants and will prevent clay movement where these ions are abundantly available in the soil solution. The eluviation–illuviation of clays is therefore at a maximum under pH conditions of moderate acidity (5.5–6.5), an acidity level associated with minimum Ca^{2+} and Al^{3+} concentrations in soil water. This soil acidity is a parameter that has been controlled by humans through agriculture for several millennia (see secton 7.2.1), and thus, human activities alter eluviation–illuviation patterns and control the progressive vertical organization of fine particles, soluble molecules and ions in the soil. These vertical redistributions are in an unstable dynamic equilibrium with natural biological mixing, which disturbs the soil's vertical organization.

7.2.3. *Anthropogenic factors*

In an anthropic landscape, various direct soil disturbances will lead to the archiving of episodes from the local land-use history. A common case is that of pits or depressions dug in the soil. Once the continual dredging/maintenance of these depressions is stopped, these features will accumulate sedimentary particles until filled. These particles, subjected to local natural and anthropogenic dynamics, originate mainly from local superficial horizons and may include plant and animal remains, as well as cultural artifacts. Such depressions can sometimes bear witness to now-truncated soils and often have unique properties (grain size texture, chemistry, structure) that affect aboveground communities.

A major anthropogenic disturbance of the world's cultivable soils, although superficial in terms of depth, concerns the impact of agricultural implements (hoes, ploughs, etc.). The effect on the surface and the superficial horizons of the soils leads to a leveling of micro- and mesotopographic landscapes and the disturbance of the superficial horizons. This anthropogenic effect will erase the inherited topography and reset the palimpsest of the microtopography of soils. Among the cultivable lands in western Europe, it is very difficult, except for mountain areas, to find areas that have not experienced past agricultural activity (Sabatini et al. 2018). In contrast, in a forest context characterized by a dense and continuous vegetation cover, the soil surface is protected/stabilized and therefore the conservation of its landforms is better.

The concept of anthrosol (IUSS Working Group WRB 2015) includes all soils that have been formed or deeply modified by human activities since the beginning of the Holocene. Given the vast quantity of soils that have been affected directly or indirectly by human activities, this definition is impractical for the pedologist.

7.2.4. *Effects of geomorphological processes on soil archives*

In the three previous sections, we described the biological, physical and anthropological archiving elements of pedological processes. Geomorphological processes also affect the soilscape and can produce excellent soil and sediment archives or …destroy them. Aeolian, fluvial and gravitational processes are the major erosional forces that affect a soil's environmental memory. Soilscapes lacking the protection of vegetation are very sensitive to ablation (Figure 7.2(b)). Consequently, biomes with little vegetation and agrosystems with denudated soils are not favorable for directly preserving soil archives.

Tracing soil archives is difficult in the case of transport by aeolian processes or by large fluvial systems because the constituents are mixed over a large area and the spatio-temporal information is lost entirely. In a small catchment or at the slope scale, soil material is often stored as pedosedimentary layers, and the properties of eroded soils can be traced. Even if the soil architecture is lost, some chemical parameters or organic botanical remains (e.g. seeds, charcoal, phytoliths) can be preserved (Figure 7.2(d)).

The former soil properties can be even better preserved when the horizons are buried under other layers of sediment (Figure 7.2(c)). In these cases, the soil parameters or constituents can be considered as fossilized (Yaalon 1971). These paleosols are no longer influenced by present-day climate, vegetation, farming, etc., and can be precisely dated. This is not the case of preserved soils that are in equilibrium with present-day conditions. The surficial soils also contain inherited properties and archives (see sections 7.2.1, 7.2.2 and 7.2.3), but these soils are often difficult to date because they are the result of a temporal continuum (Figure 7.2(a)).

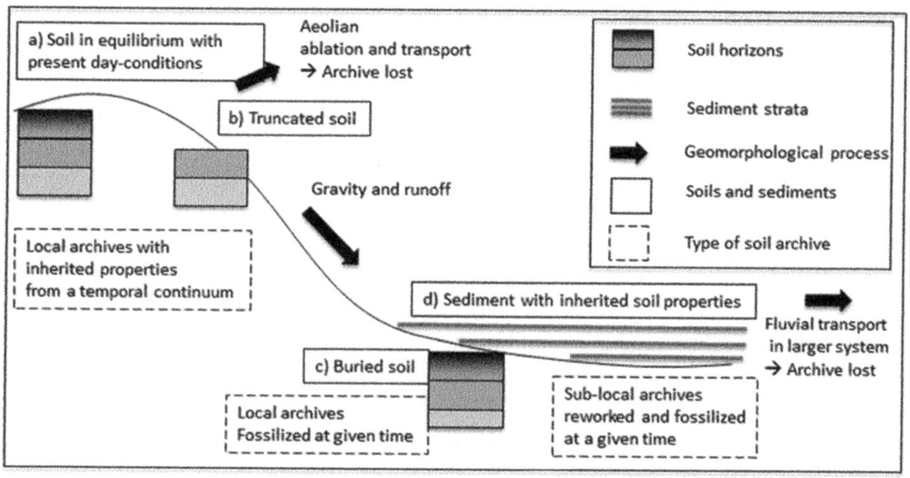

Figure 7.2. *The effects of the main geomorphological processes on the preservation and quality of soil archives. For a color version of this figure, see www.iste.co.uk/decocq/ecology.zip*

On the other hand, because they are still part of the present-day ecosystems, these soils remain relevant for historical ecology studies.

When looking for archives in soils, the main error to avoid is confusing soil horizons with soil strata. Both appear as a vertical succession of layers; however, the strata are the result of successive deposits in time that are fossilized. They can therefore be read as a relative chronology from the bottom (oldest) to the top (youngest) of the sequence. This is not the case of soil horizons that are still subjected to reorganization processes (see sections 7.1.1, 7.1.2 and 7.1.3). Consequently, no relative chronology can be read from the soil horizons, and the age or residence time of every incorporated constituent must be considered.

7.3. Examples of soil archives and their influence on current ecosystems

7.3.1. *Chemical archives, witnesses of progressive soil transformations*

The chemical parameters of soils (i.e. pH, C, N, P, exchangeable cations) are very rarely in equilibrium and are therefore subject to evolutionary dynamics at an annual to millennial scale. The inherited initial state is the starting point for these dynamics; for example, this could be a parent material newly accessible to pedogenesis. It can also be a tipping point, such as climate or land-use change. For the chemical parameters of soils, the subsequent evolutionary dynamics will tend towards a new equilibrium state. The history of a soil is therefore important for explaining the state of current chemical characteristics as well as for understanding the successions of past and future states. The associated soil biotas are subject to the influence of these chemical parameters and are likely to be structured by the achieved evolutionary trajectory.

To illustrate this point, we can look at soil pH, one of the most structuring parameters for edaphic ecosystems. Under humid climates, soil pH is subject to acidification by the leaching and/or the input of protons (from atmospheric precipitation, litter decomposition, root exudation and respiration) in the upper part of the soil (Slessarev et al. 2016). This acidity increase reduces the quality and quantity of the harvests of many food crops. To fight against this natural process of acidification, farmers, especially in northwestern Europe, have added limestone amendments for more than two millennia, thus neutralizing the acidity of the soil (see Pliny the Elder: Book XVII-Chapter 4). A spatio-temporal study of silty soils in northern France compared two chronosequences of land-use change (Figure 7.3). The authors compared plots that had been deforested and cultivated within the past few years to several centuries and plots where afforestation (on cultivated land) occurred recently to nearly 1.6 millennia ago (Brasseur et al. 2018). The soil pH of the deforested chronosequence revealed that the tilled surface horizon, initially acidic, will require a few years to approach neutrality, whereas this neutralization will take one to three centuries at more than 1-meter depth.

Conversely, when cultivated land (with neutral pH) is abandoned and afforested, the natural acidification dynamics are slower. Indeed, after 1.6 millennia of afforestation of previously cultivated land, although the pH at the surface has decreased, the pH of deeper horizons has lowered to a much lesser extent, indicating a slow process that is still underway.

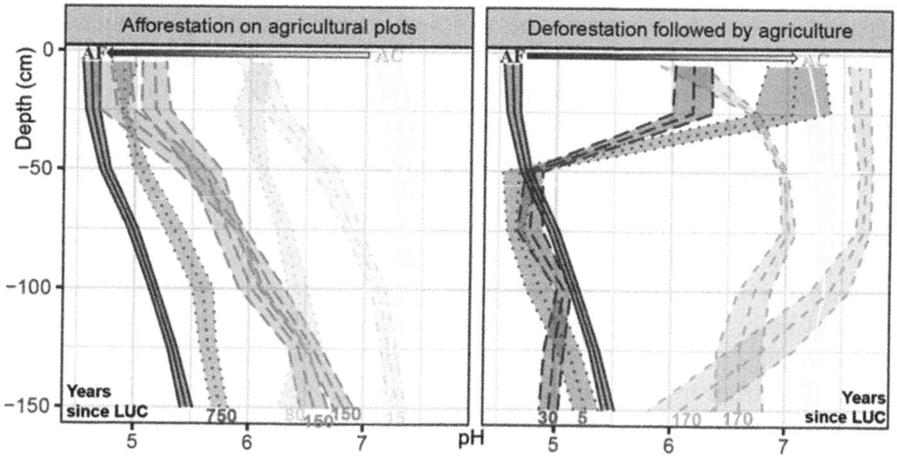

Figure 7.3. *Mean (n = 6; ± SE) pH(water) profiles in luvisols from the loess plateau in northern France characterized by differences in deforestation and afforestation ages. Left-hand graph: pH(water) profiles for agricultural plots that were afforested at various times in the past. Right-hand graph: plots deforested at various times in the past for agricultural use. AF: ancient forest (circa. 1,600 years BP), AC: ancient culture (>500 years BP), LUC: land-use change. For a color version of this figure, see www.iste.co.uk/decocq/ecology.zip*

The characterization of these evolutionary soil dynamics makes it possible to identify the successions of ongoing and forthcoming chemical states for soils in similar contexts (parental material, climate, land-use). This work also reveals that the past agricultural use in the now-forested lands of northwest Europe can still be characterized owing to this very gradual acidification of the soil. Lowland temperate forests with a primeval character no longer exist in Europe (Peterken 1996), and most European forests have been cultivated in a more or less distant past (Sabatini et al. 2018). Thus, soil pH profiles may be an indicator of afforestation oldness in many forested areas. A wealth of literature also describes the long-lasting inherited chemical effects of past human agricultural activities on soils that can be deciphered from the present-day plant community composition (e.g. Dambrine et al. 2007).

7.3.2. *Physical archives: reading the soil pit profile and microtopographic features*

An essential method for accessing the in situ physical properties of the soil archives, for example, horizonation, soil structure and disturbance, and density anomalies, is studying soil pits. This approach is the best way to describe the spatial relationships between horizons, the soil and parent material, pedofeatures (e.g. redoximorphic and textural characteristics, secondary precipitations), and biological and physical pedoturbations. Reconstructing the history of the soil then consists of unraveling the spatial relationships to derive a chronological sequence of events that have affected the soil. We can do this using simple logical principles borrowed from stratigraphy, including superposition (limited use, see sections 7.1.2 and 7.1.4), inclusion and cross-cutting. The progression in horizon individualization from the initial parent material is also a key criterium. As an example, the podzol profile (Figure 7.1) evolves through the following steps: 1) the parent material is accessible to pedogenesis, 2) acidification, horizonation and podsolization, 3) formation of dark horizontal stripes and 4) intense reworking (cross-cutting) of topsoil and eluviated horizons by the intrusion of roots and burrows following a possible local nutrient enrichment. Such enrichment can reactivate the biological activity (and bioturbation) of these very nutrient-poor soils (Robin 2005).

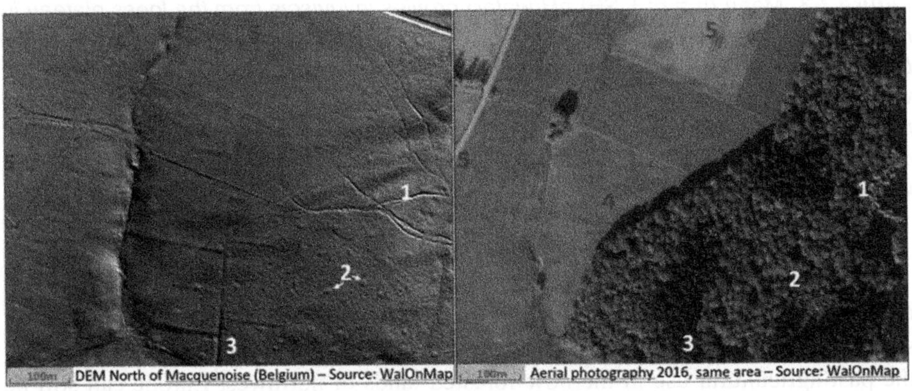

Figure 7.4. *Aerial photography and DEM of the same area with ancient forested (1, 2, 3) and deforested (4, 5; cleared ca. 150 years ago) sectors. 1) Track ways, 2) charcoal kilns, 3) demarcation of old (probably Roman) agricultural plots, 4) less visible under pasture and 5) no demarcation under culture. The darker spots are former charcoal kilns. For a color version of this figure, see www.iste.co.uk/decocq/ecology.zip*

Since the 2000s, laser remote sensing techniques have revolutionized the observation of soil microtopography. The generated digital elevation models (DEM) offer the opportunity to observe and interpret landscapes invisible to the eye, especially for land having a forest cover. In the Thiérache forest (near franco-belgian frontier), for example, DEM (Figure 7.4) has revealed numerous old agricultural parcel limits, ancient trackways and charcoal kilns. In recently deforested areas, such memories of land-use disappear gradually under pastures and much more quickly under the recurrent impact of ploughing in cultivated fields.

7.3.3. *Soil organic matter*

Soil organic matter (SOM) is a functional category of soil constituent. Its quality, chemical structure, quantity and age vary within the soil profile and the succession of horizons. Transport processes and biological and anthropological transformations (see sections 7.1.1 and 7.1.2) control these distributions. Vegetation and the microbial mass that transforms this vegetation are the main sources of OM (Kögel-Knabner 2002). The OM derives mainly from the degradation of plant fragments within the litter or from root exudates. The extraordinary biochemical diversity of the plants is partly reflected by the SOM even after transformation and microbial degradation. When researchers manage to trace and date SOM, it can serve as a proxy of past vegetation and past landscape dynamics. SOM contains thousands of different organic molecules. The classification of these complex components depends on the tools at hand and the scale of the study.

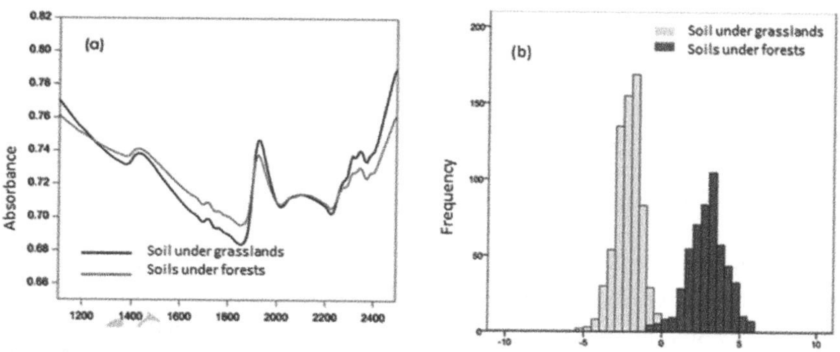

Figure 7.5. *Soil organic matter determined by near-infrared spectroscopy. (a) Near-infrared spectra of soils and (b) multiple discriminant analysis on a set of 1,333 reference spectra from 75 sites under temperate forest and grassland cover. For a color version of this figure, see www.iste.co.uk/decocq/ecology.zip*

In the last 20 years, numerous studies using gas chromatography–mass spectroscopy (GCMS) and other powerful methods have identified some very specific molecules or isotopic ratios that identify particular categories of vegetation or sometimes specific species (Trendel et al. 2010). Because they are time-consuming, these methods are often applied to only a few soil profiles and are of little use for landscape-scale studies. The fingerprinting approach of SOM by near-infrared spectroscopy is less accurate and functions as a black box in regard to SOM molecular chemistry, but it is fast and effective and provides an overall picture of the SOM. As well, the existence of an extensive catalogue of reference soil samples with a well-known ecological history (Ertlen et al. 2010; Figure 7.5) allows near-infrared spectroscopy to discriminate SOM according to its plant-association source. It is thus an effective tool for historical and landscape ecology.

7.3.4. *Botanical remains*

Soils are not well suited for the long-term preservation of organic botanical remains. Biogeochemical activities induce the degradation of organic material, limiting their retrospective representativeness. Such is the case of plant macroremains that are assimilated into the carbon cycle by organic degradation. Nonetheless, several botanical remains have specific features that make them resistant to biogeochemical degradation. Once these botanical remains are extracted from soil, taxonomic analyses of these samples can shed valuable insight into the studied organisms and provide a portrait of the environmental context (Ferguson 2005). This resistance to degradation is observed with certain microfossils such as pollen grains, owing to their exine composed of sporopollenin, or phytoliths, being microscopic structures composed of silica. These microfossils, when found in significant quantities in soils, can provide information about past land-use and vegetation history (e.g. Robin et al. 2012; Contreras et al. 2014).

These indicators are often difficult to use, however, because establishing the chronological framework is a challenge in a soil. Assessing the chronology of an archive is a key step in any retrospective study. Indeed, most soils are continuously mixed (see section 7.1.1); therefore, soils do not present a chronologically structured stratification (Carcaillet 2001). This contrasts with chrono-stratified sedimentary sequences, such as those recovered from lakes or mires, in which chronological archiving is nearly continuous from the sequence bottom to the sediment surface. Only in the case of pedosedimentary accumulation that have clear boundaries (Figure 7.2(d)) is it possible to identify a relative stratigraphy in soil archives (Robin et al. 2014). However, more precise chronological insights can be obtained from charred botanical remains. Such indicators can be preserved for a long period and are datable directly by ^{14}C measurements (see section 7.1.4). The carbonization of botanical remains, possibly related to anthropic or natural fires, transforms them into inorganic carbon remains, allowing them to resist biogeochemical degradation. Moreover, charcoal remains taxonomically identifiable, as its anatomical structures are well preserved during carbonization, except for some deformation and shrinkage (Braadbaart and Poole 2008). Quantitative and qualitative analyses of the charcoal assemblages from soils provide insight into fire and forest vegetation history (Nelle et al. 2013). Charcoal pieces must be large enough to be taxonomically identifiable (i.e. megacharcoal, >800 μm). Therefore, soil megacharcoal assemblages are relevant in fine-scale studies (Feiss et al. 2017) or at a catchment scale when, for example, studying erosive events (Robin et al. 2014).

Overall, investigations that combine diverse types of botanical remains or diverse types of indicators are the most relevant to obtain the best possible, reliable and complete picture of past environments (Robin et al. 2012; Contreras et al. 2014).

7.4. Conclusion

Soils are a complex environmental interface where the chronological succession of environments can produce distinct archives having a complex interpretation, much like studying an environmental palimpsest. In this context, it is best to combine the skills of different specialists in the (archaeo-) environmental sciences. The combined study of the different soil archives (plant remains, SOM, chemical parameters, macro- and micromorphological features, microtopography, archaeology) allows an increasingly less lacunary picture to emerge of the environmental history and its legacies in the soil; an environmental story whose heritage, while fading, still structures today's terrestrial ecosystems.

7.5. References

Braadbaart, F. and Poole, I. (2008). Morphological, chemical and physical changes during charcoalification of wood and its relevance to archaeological contexts. *J. Archeol. Sci.*, 35(9), 2434–2445.

Brasseur, B., Spicher, F., Lenoir, J., Gallet-Moron, E., Buridant, J., Horen, H. (2018). What deep-soil profiles can teach us on deep-time pH dynamics after land use change? *Land Degrad. Dev.*, 29(9), 2951–2961.

Carcaillet, C. (2001). Are Holocene wood-charcoal fragments stratified in alpine and subalpine soils? Evidence from the Alps based on AMS 14C dates. *Holocene*, 11(2), 231–242.

Contreras, D.A., Robin, V., Gonda, R., Hodara, R., Dal Corso, M., Makarewicz, C. (2014). (Before and) after the flood: A multiproxy approach to past floodplain usage in the middle Wadi el-Hasa, Jordan. *J. Arid Environ.*, 110, 30–43.

Dambrine, E., Dupouey, J.-L., Laüt, L., Humbert, L., Thinon, M., Beaufils, T., Richard, H. (2007). Present forest biodiversity patterns in France related to former roman agriculture. *Ecology*, 88(6), 1430–1439.

Ertlen, D., Schwartz, D., Trautmann, M., Webster, R., Brunet, D. (2010). Discriminating between organic matter in soil from grass and forest by near-infrared spectroscopy. *Eur. J. Soil Sci.*, 61(2), 207–216.

Feiss, T., Horen, H., Brasseur, B., Lenoir, J., Buridant, J., Decocq, G. (2017). Optimal sampling design and minimal effort for soil charcoal analyses considering the soil type and forest history. *Veg. Hist. Archeobot.*, 26(6), 627–637.

Ferguson, D.K. (2005). Plant taphonomy: Ruminations on the past, the present, and the future. *Palaios*, 20(5), 418–428.

Kögel-Knabner, I. (2002). The macromolecular organic composition of plant and microbial residues as inputs to soil organic matter. *Soil Biol. Biochem.*, 34(2), 139–162.

Lavelle, P., Spain, A., Blouin, M., Brown, G., Decaens, T., Grimaldi, M., McKey, D., Mathieu, J., Velasquez, E., Zangerlé, A. (2016). Ecosystem engineers in a self-organized soil: A review of concepts and future research questions. *Soil Sci.*, 181(3–4), 91–109.

Nelle, O., Robin, V., Talon, B. (2013). Pedoanthracology: Analysing soil charcoal to study Holocene palaeoenvironments. *Quatern. Int.*, 289, 1–4.

Peterken, G.F. (1996). *Natural Woodland: Ecology and Conservation in Northern Temperate Regions*. Cambridge University Press, Cambridge.

Robin, A.-M. (2005). Épisodes majeurs de la podzolisation en forêt de Fontainebleau (France). Essai de synthèse à l'aide du radiocarbone. *C.R. Geosci.*, 337(6), 599–608.

Robin, V., Rickert, B.-H., Nadeau, M.-J., Nelle, O. (2012). Assessing Holocene vegetation and fire history by a multiproxy approach: The case of stodthagen forest (northern Germany). *Holocene*, 22(3), 337–346.

Robin, V., Bork, H.-R., Nadeau, M.-J., Nelle, O. (2014). Fire and forest history of central European low mountain forest sites based on soil charcoal analysis: The case of the eastern Harz. *Holocene*, 24(1), 35–47.

Sabatini, F.M., Burrascano, S., Keeton, W.S., Levers, C., Lindner, M., Pötzschner F., Verkerk P.J., Bauhus J., Buchwald E., Chaskovsky O. et al. (2018). Where are Europe's last primary forests? *Divers. Distrib.*, 24(10), 1426–1439.

Šamonil, P., Král, K., Hort, L. (2010). The role of tree uprooting in soil formation: A critical literature review. *Geoderma*, 157(3), 65–79.

Slessarev, E.W., Lin, Y., Bingham, N.L., Johnson, J.E., Dai, Y., Schimel, J.P., Chadwick, O.A. (2016). Water balance creates a threshold in soil pH at the global scale. *Nature*, 540(7634), 567–569.

Stinchcomb, G.E., Driese, S.G., Nordt, L.C., DiPietro, L.M., Messner, T.C. (2014). Early Holocene soil cryoturbation in northeastern USA: Implications for archaeological site formation. *Quatern. Int.*, 342, 186–198.

Trendel, J.M., Schaeffer, P., Adam, P., Ertlen, D., Schwartz, D. (2010). Molecular characterisation of soil surface horizons with different vegetation in the Vosges massif (France). *Org. Geochem.*, 41(9), 1036–1039.

Yaalon, D.H. (1971). Soil-forming processes in time and space. In *Paleopedology. Origin, Nature and Dating of Paleosols,* Yaalon, D.H. (ed.). Israel University Press, Jerusalem.

8

Continuous and Nested Time in Historical Ecology: Application to Soil Studies

Damien ERTLEN
University of Strasbourg, France

8.1. Interdisciplinarity and time in historical ecology

Historical ecology (HE) is about interdisciplinarity (Szabó 2015). Anthropology, archaeology, ecology, ethnography, geography, geology and history are the disciplines most commonly found within the framework of HE. The combination and relative importance of these disciplines in HE vary somewhat between authors and depend on how these disciplines are defined and delimited. Moreover, HE can be considered as "deep interdisciplinarity" because it is not simply a temporary collaboration or a discipline assisting another by providing a different method or having a different point of view. Indeed, the researchers involved in HE share a common set of questions that require addressing in a research project (Balée 2006). Studying the dynamic relations over time between humans, "our species", and nature, "our environment", is one means of formulating the central research question (Balée 2006; Crumley 2007; Szabó 2015). This human–environment interaction is also often considered as a central question for geographers (Dauphiné 2003). The assumption that all environments on Earth have been affected by humans (Balée 2006; Rostain 2016) is widely accepted by HE researchers. This idea provides an overarching dimension to any project.

Several disciplines involved in HE deal intrinsically with time: geology through the Earth's history, biology via the history of life and archaeology and history involving the history of societies. Time is a common and relevant topic for practicing interdisciplinarity. All sciences that focus on the environment eventually became historical sciences, as we cannot understand the present without understanding the past (Legeay 2000). This position is increasingly accepted, particularly in ecology, because time coupled with space represents one of the fundamental axes shaping ecological processes (Wolkovich et al. 2014). White (2007), for example, demonstrated the necessity of considering time and various timescales in landscape ecology and landscape management/restoration. She explains how a lack of historical perspective or the use of an improper timescale can lead to a misunderstanding of landscape dynamics. Crumley (2007) explains how failing to consider time led some American ecologists to an impasse, as they continued to look for "pristine" ecosystems lacking a human influence. In a similar vein, Erickson (2020) describes how the "discoverers" of pristine forest in Amazonia are very often frustrated when visiting the forest together with archaeologists, who quickly find evidence of a past human presence. The emergence of climate change as an issue has further challenged researchers across multiple disciplines to better understand processes over a longer temporal scale.

The author of this chapter is a geographer … talking about time. Indeed, work on dynamic processes (social or natural) involves more fields than the traditional disciplines that study the past. The lack of a strong temporal framework in geography likely makes it easier for geographers to switch from one timescale to another, similar to how geographers traditionally act in regard to spatial scales. Timescale issues and errors obtained when using the inappropriate timescale have been previously identified by Rymer (1979). Time in HE has been conceptualized by Sinclair et al. (2018) who placed a specific focus on complex systems and the necessity to move beyond the linear, Newtonian definition of time. Here, I focus on two practical means for approaching time with the goal of improving communication between the practitioners of HE. To provide an example of a study tool, I rely on the use of soils and their temporal organization in HE studies.

8.2. Continuous time

Continuous time is used by historians and the historical sciences. It is also the vision of time learned at school. For the geologist, the Earth's historian, it begins with the Big Bang (ca. 4.5 billion years BP). Time is then divided by a hierarchical and universal system into aeons, eras, periods, epochs and stages (Figure 8.1). The limits between these time entities are often linked to life on Earth; for example, the start of the second aeon, the Archean, is associated with the first record of life on Earth. The Cretaceous is associated with the appearance of flowering plants. The Quaternary, until 2009, was a period associated with the appearance of Hominidae; however, following the discovery of older hominids, the Quaternary was downgraded to an epoch, forming part of the Cenozoic era. Other transitions between geological subdivisions are linked to changes in atmospheric chemistry and geological events.

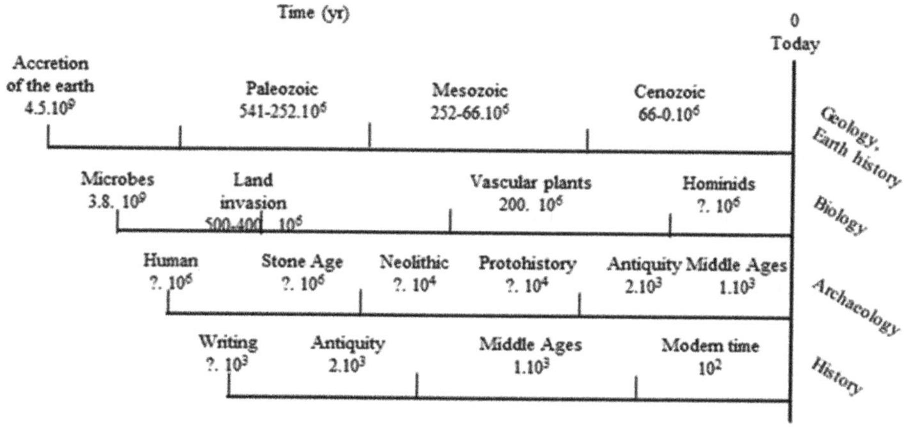

Figure 8.1. *History of human–environment relations: a continuous-time perspective*

Similarly, historians divide time into a series of successive periods. This system is not as universal as that for geology; however, traditional periods, such as prehistory, the Neolithic Age, antiquity and the Middle Ages, are widespread and globally well understood, despite regional differences in terms of the absolute limits of these periods. This subdivision of time serves not only as a tool to facilitate the organization of Earth or human history, but often it also serves to organize university departments and research projects. Consequently, specialists of one period (one slice in time) develop their own community, which can sometimes be disconnected from other communities. Each community develops its own methods, and these closely knit research spheres are often disconnected from other periods of time – even including the present. This means of dividing time also adds to the difficulty for interdisciplinary communication, given that the timescales of interest for each discipline do not necessarily match and are often difficult to bridge (Figure 8.1). Crumley (2007) underlines the challenge in bringing together these different systems within an "integrated framework" dedicated to HE. On the other hand, these chronology-based divisions cannot be neglected because they remain the most common temporal divisions used by HE researchers.

8.3. Nested time

An alternative way to divide time is to consider nested portions of time rather than continuous time slices. We can use the image of Russian dolls, where each doll represents a different timescale. The above-mentioned geological divisions can also be viewed as a nested timescale system because it is a hierarchical structure. In the approach described in this chapter, however, all timescales end (or begin) at the same point, which is the present ("Today" in Figure 8.2). This point is important because although some archives such as sediments may have stopped accumulating information at an earlier moment in time, we nonetheless study these archives in the present and must consider the amount of time that the archive has been preserved and the degradation of this archive over time. This view agrees with one of the primary objectives of HE, which is to explain present-day landscapes and ecosystems from the viewpoint of these multiscale temporal and spatial dynamics. The French historian Braudel (1980), developing short- and long-term history, provided a precursor to this nested time approach; however, it was limited to a centennial timescale.

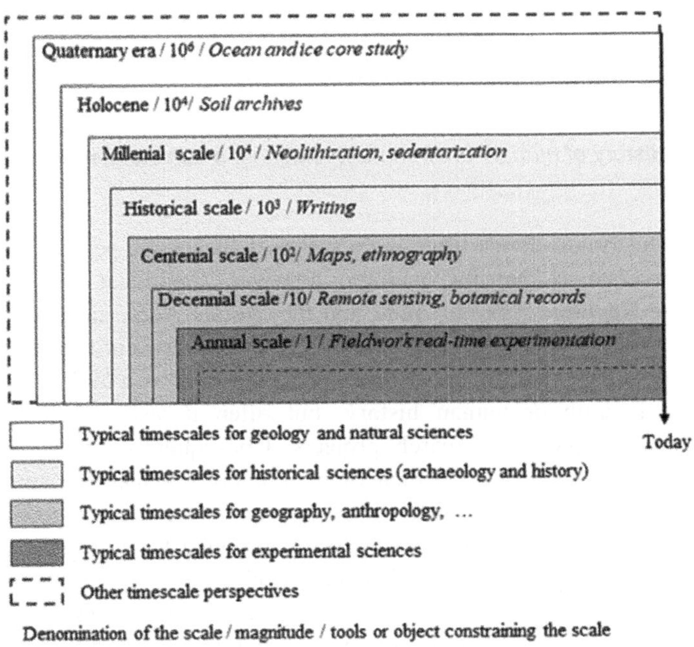

Figure 8.2. *History of human–environment relations: a nested-time perspective*

We could view very short timescales, at an hourly or daily scale, as the smallest scales of interest (smallest doll); in practice, however, these very short-term scales are rarely used in HE. Data at a yearly or decadal scale are much more common in HE and are described as short-term history. (In practice, use of these short-term and long-term denominations alone is quite challenging because their magnitude can vary between disciplines and researchers.)

Centuries and millennia are also appropriate timescales in studies of ecological history. We can also ask what is the largest doll and how far back does HE reach into the past. Szabó (2015) distinguishes three forms of past: the past *sensu lato*, the human past and that part of the human past for which written records are available. A consensus is emerging that HE is the history of the human–environment relationship; therefore, it does not become necessary to extend beyond this time frame. Some older elements, however, particularly from geology or biology, are often useful for HE, not to study their own particular histories per se, but because they are inherited and are implicated in the human–environmental interactions. Crumley (2007) defends the idea that temporal ranges are limited only by what data are available and thus relevant timescales for a study can range from very recent to geological timescales. The development of agriculture has also been viewed as a key point in human history and sometimes as a starting point for HE. Consequently, post-Neolithic, or more generally the Holocene, has been considered as the time frame for HE. However, the controlled use of fire and other forms of plant management have been described as occurring long before crop cultivation and neolithization.

The framework of HE provides an intellectual "contact zone" between disciplines (Pratt 1991 in Meyer and Crumley 2012). To be effective, the contact must concern an object of study occurring at the same time or with overlapping timescales. With nested time, the different timescales always have a minimum area of overlap. Describing this nested system is relatively simple using an ordering of chronological scales, although in practice, HE researchers have various strategies and constraints when selecting the appropriate scales for study. In the following sections, we describe these practices and emphasize the interest in combining different timescales to strengthen the conclusions of HE studies.

8.4. Different disciplines, different tools

Even if there were to be an eventual agreement among researchers within the interdisciplinary consortium of HE on this necessity of using a nested time approach, researchers would still have to agree on the magnitude and limits of these temporal scales. Clearly, the temporal magnitude varies among disciplines. In some cases, as described for continuous time, this is often a question of habit or discipline-specific methods; it is not difficult to convince colleagues to switch timescales when we demonstrate the advantages of the novel method, the relevance of the new scales and the attraction of newly available data. In most cases, however, it is not simply a question of methodology, but rather this method being determined by the tool. Nearly all the tools are restricted in their magnitude of time, and below, we provide three examples to highlight this point.

(i) Old maps and old aerial photographs are highly useful for reconstructing past environments and landscapes. Geographers and historians are quite familiar with the use of such tools. Increasingly, map and photograph collections are being scanned, digitized and made widely accessible within databases. They often provide clear and precise information in relation to the spatial distribution of forests, grasslands, agricultural plots, etc. These tools commonly provide detailed information about vegetation type. In Europe, maps from the 18th century begin to provide sufficiently precise and reliable information. From the mid-19th century, maps become detailed, clear and easy to digitize. By the 20th century, maps – coupled with the arrival of aerial photographs – are increasingly available, and the more frequent surveys of a given land parcel make it possible to reconstruct landscape dynamics at a higher spatio-temporal resolution. Maps produced prior to the 18th century are available; however, they are difficult to align, the information contained within is often unreliable and the mapped areas are dispersed in space and time. Consequently, the interest in using old maps markedly falls when changing from the timescale of the last three centuries to a millennial timescale. In this latter case, a historian or a paleoecologist must rely on other better-adapted tools to access older archives at this longer timescale.

(ii) The Neolithic is generally now rejected as a starting point for HE. Nevertheless, the social changes and, in particular, the arrival of a novel sedentary lifestyle produced a massive change in the quantity and readability of archaeological remains. Consequently, there is a wide methodological gap between pre- and post-Neolithic sites in terms of their excavation and study. This gap also involves temporal organization and time resolution; for example, the relative chronology built from Neolithic pottery can approach a time resolution of a single human generation (30 years), a very high resolution for the mid-Holocene. The chronologies of earlier periods are based on a few pieces of stone tools and scattered radiocarbon dates. In contrast, the material available in the Neolithic, both for relative and absolute chronologies, is much more abundant. Therefore, the dating of sediments by paleoecologists is greatly more critical prior to the Neolithic. A collaboration between a paleoecologist and an archaeologist can offer completely different experiences depending on the archaeologist's temporal speciality, for example, Neolithic versus Paleolithic.

(iii) The soil archives presented by Brasseur et al. (see Chapter 13) and also discussed below have time limits that vary according to their location. In temperate Europe, for example, it is accepted that all soils are younger than the last deglaciation. During the last glacial period, which attained a maximum extent 20,000 years BP, soils were destroyed either directly by ice sheets and glaciers over northern Europe or by periglacial erosional processes occurring in the periphery of the ice. Consequently, a European soil's environmental memory is restricted to the post-glacial period. Soil scientists and paleoecologists can jump to an earlier timescale by focusing on aeolian sediments or relict paleosoils, although this requires a very different approach.

8.5. Examples of nested and continuous time: soils and strata

Soil lies at the earth–atmosphere interface and is studied by soil scientists (pedologists). Soils are commonly used in HE because they store information about past environments (i.e. Koerner et al. 1997). They are complex and open systems under the influence of geochemical and biological processes, and soils are often misunderstood by researchers from other disciplines (Schwartz 2012). One major source of confusion occurs between the notions of strata, which are the elementary bodies of sedimentological stratigraphy, and horizons, which are the elemental bodies of soil. A major difference between these two types of layers, which sometimes have a similar appearance to inexperienced observers, is the organization of time.

The organization of strata is well known and follows the model of continuous time. Usually, time is organized vertically from oldest at the bottom to most recent at the top, and therefore we can talk about superimposed time. Geologists and archaeologists are very familiar with this concept; however, the different visions of time between the disciplines and the different applied timescales are sometimes problematic. It requires great effort to clearly understand timescales used by other disciplines. The main issue arises when comparing absolute time and relative time. On the one hand, the expanded development of radiocarbon dating and other absolute-age dating methods, such as optically stimulated luminescence (OSL) and in situ cosmogenic nuclides (i.e. ^{10}Be, ^{26}Al), has produced a greater abundance of absolute dates from natural sediments and has improved the accuracy of the absolute scale. On the other hand, the archaeological findings and typologies are also very accurate and highly relevant when they are compiled into large, spatialized databases. Matching up absolute time and relative time remains the main challenge.

Soils and soil horizons follow very different time organizations than strata. They are the result of the redistribution and reorganization of elements through biological and physical processes. If the soil remains unburied, these processes continue. These still-functioning soils are appealing for HE studies because the soil cover is (i) nearly continuous across the surface and (ii) most information contained within the soil is local, thereby allowing the comparison of, at a specific spatial scale, vegetation, archaeological features and other objects of interest for HE. The temporal organization in these living soils corresponds to the nested organization of time.

Depth in soil	Chemical compound	Magnitude/timescale
Top of soil	Leaves and other vegetation debris	Years
Upper 20 cm/surficial horizon	Total organic matter	Decades
	Lignin, bacterial hexosamines, glucosamines*	Decades
	n-alkanes, saccharides*	Century
Deep horizons	Total organic matter	Centuries to millennia
	n-alkanes	Millennia
	Glycerol dialkyl glycerol tetraethers (GDGTs)	Millennia

*According to Schmidt et al. (2008).

Table 8.1. *Organic matter components and their respective transit times*

Some elements accumulate in soils; however, most elements are in transit in soils, and this transit time differs greatly among elements. The most resistant mineral grains, such as quartz, can remain in soils for several million years before being weathered. In contrast, some organic components remain in the soil for only a single season or a few days. This transit time varies greatly, even by an order of magnitude, among the various types of organic molecules (Table 8.1). Radiocarbon dating of soil organic matter, an approach commonly used by HE researchers, has been important in soil studies since the 1980s (Becker-Heidmann and Scharpenseel, 1992). Because this organic matter is heterochronous, it is exceedingly difficult to determine its absolute age. Rather, soil scientists determine a mean residence time (MRT, Figure 8.3) for this organic matter. This MRT depends on depth, soil type and multiple interdependent natural variables, including temperature, pH, microbial activity and vegetation type. MRT is also influenced by soil management strategies and human-related soil disturbance, thereby making MRT extremely relevant in HE.

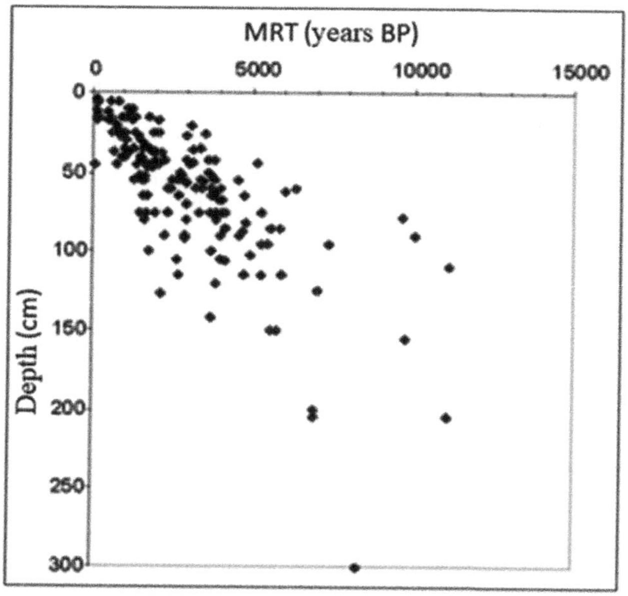

Figure 8.3. *Total soil organic matter mean residence time (MRT) from a large set of soils that includes various soil types under different environmental conditions*

8.6. Conclusion

Historical ecology, as a science using the past to understand the present, cannot avoid questioning time and timescales. Disciplines differ in their approaches for dividing time. Here, we propose synthesizing these various means into two main approaches: continuous and nested time. Although both approaches have their unique advantages and limitations, nested time is best adapted for building a strong and comprehensive dialogue between the disciplines working in HE. Nevertheless, this approach does not eliminate the interdisciplinary differences and most researchers remain narrow specialists within a single discipline. For constructive dialogue within the framework of HE, however, we recommend that researchers pay particular attention to the chronological views of closely related sciences to develop their own knowledge of these alternative visions of time.

8.7. References

Balée, W. (2006). The research program of historical ecology. *Annu. Rev. Anthropol.*, 35, 75–98.

Becker-Heidmann, P. and Scharpenseel, H.W. (1992). Studies of soil organic matter dynamics using natural carbon isotopes. *Sci. Tot. Env.*, 117–118, 305–312.

Braudel, F. (1980). *On History*. University of Chicago Press, Chicago.

Crumley, C.L. (2007). Historical ecology: Integrated thinking at multiple temporal and spatial scales. In *Issues and Concepts in Historical Ecology: The Past and Future of Landscapes and Regions*, Crumley, C., Lennartsson, T., Westin, A. (eds). Cambridge University Press, Cambridge.

Dauphiné, A. (2003). *Les théories de la complexité chez les géographes*. Anthropos, Paris.

Erickson, C.L. (2020). Foreword. In *Methods in Historical Ecology: Insights from Amazonia*, Odonne, G. and Molino, J.-F. (eds). Routledge, London.

Koerner, W., Dupouey, J.-L., Dambrine, E., Benoît, M. (1997). Influence of past land use on the vegetation and soils of present day forest in the Vosges mountains. *J. Ecol.*, 85, 351–335.

Legeay, J.-M. (2000). Les temps de l'environnement. In *Les temps de l'environnement*, Barrué Pastor, M. and Bertrand, G. (eds). Presses Universitaires du Mirail, Toulouse.

Meyer, W.J. and Crumley, C.L (2012). Historical ecology: Using what works to cross the divide. In *Atlantic Europe in the First Millennium BC: Crossing the Divide*, Moore, T. and Armada, X.L. (eds). Oxford Scholarship Online, Oxford.

Rostain, S. (2016). *Amazonie : un jardin sauvage ou une forêt domestiquée*. Actes Sud/Errance, Paris.

Rymer, L. (1979). Historical ecology and environmental conservation. *Env. Conserv.*, 6, 199–200.

Schmidt, M.W.I., Torn, M.S., Abiven, S., Dittmar, T., Guggenberger, G., Janssens, I.A., Kleber, M., Kögel-knabner, I., Lehmann, J., Manning, D.A. et al. (2008). Persistence of soil organic matter as an ecosystem property. *Nature*, 478, 49–56.

Schwartz, D. (2012). Les temps du sol : interprétations temporelles de l'archivage pédologique dans les approches paléoenvironnementalistes et géoarchéologiques. *Etude et gestion des sols*, 19, 50–66.

Sinclair, P., Moen, J., Crumley, C.L. (2018) Historical ecology and the longue durée. In *Issues and Concepts in Historical Ecology*, Crumley C.L., Lennartsson, T., Westin, A. (eds). Cambridge University Press, Cambridge.

Szabó, P. (2015). Historical ecology: Past, present and future. *Biol. Rev.*, 90(4), 997–1014.

White, K.A. (2007). Invoking the ghosts of landscapes past to understand the landscape ecology of the present… and the future. In *Issues and Concepts in Historical Ecology: The Past and Future of Landscapes and Regions*, Crumley, C., Lennartsson, T., Westin, A., (eds). Cambridge University Press, Cambridge.

Wolkovich, E.M., Cook, B.I., Davies, T.J. (2014). Progress towards an interdisciplinary science of plant phenology: Building predictions across space, time and species diversity. *New Phytol.*, 201(4), 1156–1162.

9

The Analysis of Relic Charcoal Kilns for the Assessment of Forest Trajectories

Vincent ROBIN[1], Alexa DUFRAISSE[2] and Claudia OLIVEIRA[1]

[1] *University of Lorraine, Metz, France*

[2] *Muséum National d'Histoire Naturelle, Paris, France*

9.1. Introduction

The demand for a reliable supply of energy to meet industrial developments has continuously increased from the Bronze Age onwards. Energy was supplied primarily from charcoal produced after woodland biomass exploitation until the rise of fossil energy use during the 19th century (Pain 2017). Such biomass exploitation for charcoal production represented an anthropogenic disturbance to the dynamics of the forest ecosystems. Therefore, historical charcoal production has had a significant influence on past forest trajectories, from punctual and local, to large and long lasting, with, moreover, possible heritages on the present-day state of forests (Ludemann 2010; Bonhage et al. 2020; Máliš et al. 2020). Thus, the reconstruction of historical charcoal production regimes (e.g. magnitude, intensity, frequency, etc.) might provide key insights to improve our understanding and knowledge about the historical ecology of forest ecosystems and how it has been used as resource corollary of human development.

Valuable data can be obtained from specific archives of past charcoal production that are the remaining charcoal production structures (i.e. kilns) and charcoal pieces

Historical Ecology,
coordinated by Guillaume DECOCQ.

remains (Nelle 2003; Gocel-Chalté et al. 2020). The carbonization of woodpiles (Figure 9.1a) in soil pits during antiquity and in mound-shaped kilns from the Middle Ages onwards (Figure 9.1b) implies that there has been an on-site accumulation of ash and charcoal pieces. These remains, mixed with on-site burnt soil, constitute a charcoal-rich layer (Figure 9.1c), which represents an archive that can be used to investigate historical forest ecology. However, pit kilns are hardly identifiable in modern landscapes without specific excavation (Deforce et al. 2021). Therefore, investigations of charcoal-rich layers are conducted primarily on mound-shaped kilns. These are commonly pointed out across the globe, largely because they are fairly easy to identify from the relict platform on which the mound-shaped kilns were originally set up (Figure 9.1d).

Figure 9.1. *Various states of mound-shaped charcoal kilns: setting up the woodpile mound (a); during carbonization (b); relict charcoal-rich layer view along a longitudinal soil profile (c) and ongoing charcoal sampling of a relict charcoal kiln platform (d). For a color version of this figure, see www.iste.co.uk/decocq/ecology.zip*

9.2. Looking at the platform of the kiln

9.2.1. *Looking at the dimensions of the kiln platforms*

An analysis of the dimensions of charcoal-rich layers can provide information about past charcoal production.

First, the thickness of the charcoal-rich layer can be assessed. We can hypothesize that the thickness of the charcoal-rich layer is related to possible variations in the quantity or duration of charcoal production. However, to date, no relationship has been identified between the production processes and the thickness of the charcoal-rich layer. Even when high-resolution indicators are used to investigate the charcoal-rich layer, such as micro-morphological features and paleo-temperature records (Dupin 2018), the charcoal layer appears to be a homogeneous layer of varying thickness. The micro-morphological analyses of some charcoal-rich layers have revealed that the charcoal makers reworked the surface of the platforms between carbonization of the woodpiles (Gebhardt 2007; Dupin et al. 2017). This probably caused mixing of the accumulated materials with in situ soil, which homogenized the charcoal-rich layer. This is also why the practice of reusing kiln platforms, potentially over a long period with many cycles of production, is barely identifiable and quantifiable. This is a crucial issue when attempting to quantitatively investigate the past exploitation of forest biomass for charcoal production.

Second, the dimensions of the charcoal-rich layers can be assessed by the characterization of their surface and shape. This is achieved by on-site measurements of the minimum and maximum diameters of relict platforms. Such dimensional analysis has shown that the platforms created during the 19th and 20th centuries tend to be more ovoid in shape and larger (up to ca. 200 m^2) than the older ones, which are more circular in shape and smaller (ca. 40–80 m^2; Carrari et al. 2017; Dupin 2018; Gocel-Chalté et al. 2020). However, the surface and shape of the platforms at the regional scale vary widely and it has not been possible to identify a significant relationship between the surface and shape of the charcoal-rich layer and the age or quantity of charcoal production.

9.2.2. *Platform inventory*

The quantification of the kiln platform and the characterization of its spatial distribution might provide valuable information that could identify the historical patterns for charcoal production and its possible impact on past and present forest states (Ludemann 2010). Indeed, kiln frequency differences in space over a defined area might reflect various intensities of biomass exploitation. The exploitation might be related to specific ecosystem attributes, such as slope, or it may be related to the presence of local industries (Gocel-Chalté et al. 2020; Schneider et al. 2020). Various indicators might be used in the spatial analysis of kiln distribution, such as the distribution structure through space, neighboring distance analysis and the topographical wetness index. However, all of them rely on an inventory of kilns that should be as accurate as possible.

At the forest scale, the kilns can be inventoried through field prospections because kiln platforms are identifiable as specific landforms, especially in an area with a marked relief (Nelle 2003; Schneider et al. 2020; Figure 9.1d). However, kiln platforms are much more difficult to detect on plains because of the lack of relief within a flat landform or in areas with dense vegetation cover. Numerical tools for kiln detection might be of great importance in such cases. These numerical tools can also be used to proceed an inventory of kilns at large spatial scales. First, kilns might appear on aerial pictures as dark dots in freshly plowed crop fields. Moreover, during the last 20 years, airborne LIght Detection And Ranging (i.e. LiDAR) has emerged as a powerful tool for the detection and inventory of kiln platforms among other anthropic or natural landforms (Ludemann 2012; Puech et al. 2012). The LiDAR technique is a high-resolution "laser scanning process" that provides a set of contact points with x, y, z coordinates. These clouds of contact points can be used to derive different digital relief models that can highlight micro-topographical features, including kiln platforms (Mayoral et al. 2017; Žutautas 2017; Figure 9.2). Furthermore, the development of automatic detection tools based on deep learning processes over the last decade (Schneider et al. 2015; Trier et al. 2016) has provided powerful tools that can inventory kiln platforms efficiently at large spatial scales and extrapolate the locally gathered kiln data (i.e. upscaling, Ludemann and Nelle 2017).

Figure 9.2. *Digital elevation model derived from LiDAR treatments of the same study area across the northern Vosges in France. It highlights the presence of kiln platforms (little round forms distributed all over the pictures). Slope model (a), light exposure model (b) and Topographic Position Index model (c)*

Moreover, it is important to stress that using numerical tools to inventory kilns reduces any under-estimation of kilns which may be observed when kilns are inventoried using only field surveys. For example, in a recent study across the northern Vosges in France (Gocel-Chalté et al. 2020), both field and LiDAR inventories were independently created. A comparison between the two revealed that field inventories under-estimate the kiln frequency by about 30%, while the LIDAR inventory valid detection rate for kiln platforms was 96%.

9.3. Looking at the charcoal pieces

9.3.1. *Sampling*

Besides the dimensions of the kiln platform, the charcoal remains may also provide key insights. First, charcoal pieces from the charcoal-rich layer are sampled. For one kiln, the sampling consists of randomly collecting charcoal pieces from various depths in the charcoal-rich layer (Figure 9.1c) on manually opening small pits (e.g. 40/40 cm). The sampling can be done by on-site handpicking or by taking a volume of sediment from the charcoal-rich layer and sieving it to extract the charcoal pieces. At least 100 charcoal pieces per kiln are usually collected for the dendro-anthracological analyses. To avoid the risk of falsely recording over-represented taxa, the sampling must be done at different locations on the kiln platform (Nelle 2003). Classically, samples are taken in the center of the kiln platform and on the downstream side of the platform, which is usually richer in charcoal than the kiln itself.

For a given area under investigation, the number of kilns that must be sampled to achieve an appropriate level of representativeness depends on 1) the surface of the area being investigated and 2) the frequency of kilns in the area being investigated. In accordance with these parameters, a sampling strategy should be defined that is based on the best compromise between the expected representativeness level of the data and the possible analytical effort.

9.3.2. *Taxonomic identification*

The taxonomic identification method consists of a microscopic analysis of the sampled charcoal assemblages (larger than 800 μm/1 mm) in each studied kiln. The analysis relies on the fact that wood has specific microscopic anatomical structures that are preserved after carbonization, and which correspond to specific taxa or groups of taxa. Their identification provides the taxonomic spectrum of the ligneous species that correspond to the sampled charcoal remains. The results of the taxonomic identification are usually expressed as the relative values of the identified taxa frequency in the 100 charcoal pieces sampled and identified for each investigated kiln (Figure 9.3).

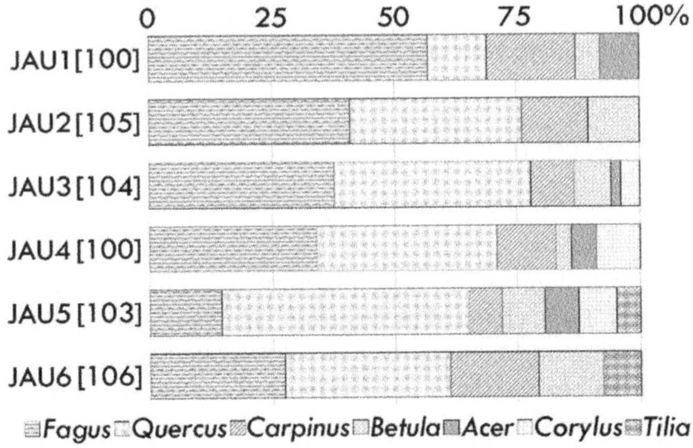

Figure 9.3. *Results of the taxonomical analysis of charcoal pieces assemblages from six kiln platforms in Meuse, France (JAU 1 to 6). The results show the proportion of identified taxa when all the analyzed charcoal fragments per kiln platform are considered. For a color version of this figure, see www.iste.co.uk/decocq/ecology.zip*

The taxonomical identification of charcoal assemblages from kilns provides information about the taxa that were available locally as a resource and the composition of the forest surrounding the investigated kilns. This is because the consumed wood for charcoal production would have been taken from the area surrounding the kiln and was dependent on the topography and the availability of local biomass (Knapp et al. 2013; Dupin 2018). Thus, specific species selection for the charcoal production is rare, even if hardwood of strongly re-sprouting tree species are usually the preferential resource for charcoal production (Grenouillet-Paradis 2012). It has been shown that charcoal can be largely produced from softwood when this is the only available resource (Deforce et al. 2013). It has been also shown that the kiln charcoal assemblages at the catchment scale strongly reflect the potential forest tree distribution, which is related to environmental conditions (Gocel-Chalté et al. 2020). For example, the taxonomic information about the charcoal assemblages might provide evidence of forest stand disturbance because light-demanding pioneer tree species were recorded in the sample (e.g. *Betula*, *Pinus*; Larsen et al. 2016). Such information about past forest dynamics becomes even more interesting when combined with structural reconstructions of the harvested forest stand and related past forest management regimes (e.g. Deforce and Haneca 2015).

9.3.3. *Dendro-anthracology*

Dendro-anthracology aims to gather the information contained in tree growth rings. Growth rings are the concentric annual rings observable on transverse sections of wood. Expressed as a function of time, tree-ring widths reflect growth dynamics, which depend on intrinsic characteristics (e.g. age, tree parts, etc.), environmental conditions (e.g. climatic events) and anthropic activity, such as forestry practices/management (e.g. coppicing).

Dendro-anthracological measurements and data acquisition rely on the same principles as dendrochronology, with some adaption because charcoal pieces are incomplete, usually contain short growth rings series and come from unknown parts of the trees (Dufraisse 2006). Thus, the application of dendro-anthracological tools requires a minimum transverse plane size of about 4 × 4 mm and at least one whole growth ring. Furthermore, to be reliable, the dendro-anthracological analyses must be done on at least 50–100 charcoal pieces per kiln. The dendro-anthracology is based on qualitative and/or quantitative observations of specific parameters, such as heartwood/sapwood discrimination, the presence of truncated tree rings and wood shrinkage during carbonization. The charcoal–pith distances also need to be measured, which can be recorded in different methodological ways (i.e. tree-ring boundaries, printed target and trigonometric calculation, and image analysis; Ludemann and Nelle 2002; Dufraisse et al. 2020). The observation of these previous dendro-anthracological parameters in current wood stands using dendrochronology means that an interpretative reference grid can be established (i.e. an anthraco-typological key; Dufraisse et al. 2018) for each tree species. This grid is necessary to classify the charcoal pieces according to their combined dendro-anthracological characteristics. Overall, the insights gathered from dendro-anthracology contribute to a more informed understanding of the biomass harvest process (e.g. type of wood, season of harvest, storage). For example, in deciduous oaks, heartwood formation takes place in 20–25-year-old trees. Therefore, the significant recording of heartwood in charcoal pieces from kilns provides a solid estimation of the minimum age of the exploited trees for charcoal production and it can be linked to forest management practices, such as coppicing (Deforce and Haneca 2015).

9.4. Looking at the ages

Chronological data recording is a key step in the analysis of historical charcoal production and its past consequences on forest ecosystems. It is possible to use different chronological markers to obtain charcoal production dates.

The most frequently used chronological marker is the measurement of radiocarbon ages (Walker 2005). To date a charcoal kiln, one or several charcoal fragments must be selected according to 1) the focus of the investigation, 2) the gathered taxonomic spectrum at the kiln scale and 3) the gathered taxonomic spectrum at the study area scale. It is important to select a fragment of charcoal that allows the age of the carbonization to be dated. Indeed, long-living trees, such as oaks of beech trees, might induce a bias in age measurement due to a potential difference of several centuries between the tree rings that were formed earlier in the life of the tree and those that formed shortly before the tree was harvested for charcoal production. Those latter rings are the ones targeted to date the carbonization. To minimize such possible "old-wood effect", charcoal pieces from short-lived tree species, such as birch or hazel, should be preferentially selected. To obtain a relevant radiocarbon date, fragments of charcoal with marked tree-ring curviness or/and with clearly visible external rings and/or sapwood can also be selected. In addition, several charcoal pieces should be used to date a kiln platform. This will provide a more reliable age distribution for the use of the platform. At larger spatial scales, such as the catchment or territory scale, a set of radiocarbon dates from different kilns should be used when assessing the chronological patterns for charcoal production. These patterns can be obtained using probabilistic models of combined radiocarbon dates, according to their calibration ranges (e.g. KDE model; Bronk-Ramsey 2016).

Recently, in addition to radiocarbon dates, the use of optically stimulated luminescence dates to identify the ages of charcoal platforms has been assessed. The technique provides interesting and promising results that enable the age range obtained by radiocarbon dating to be shortened (Karimi Moayed et al. 2020). Furthermore, it is possible to evaluate the chronological patterns for charcoal production based on the dendro-anthracological measurements. In fact, some charcoal pieces from kilns have sufficiently visible tree rings to provide dateable growth patterns that are comparable to those used in dendrochronology (Blondel et al. 2018). Finally, the chronological patterns for charcoal production may be identified by an interdisciplinary study that includes the analysis of historical written sources, such as historical maps or forest harvest registers.

Overall, such chronological information will provide important insights into the assessment of forest disturbance due to forest biomass exploitation for charcoal production.

9.5. Conclusion

Charcoal kiln analysis provides phyto-historical insights that enable researchers to reconstruct the history of charcoal production and its related consequences on ecosystems, from local to large spatial scales. Thus, the charcoal analysis of kilns can provide information about previous forest disturbances caused by biomass exploitation for charcoal production. Indeed, the charcoal analysis of kilns can be used to identify disturbance regime parameters, such as type of fuel, age of the disturbance events, magnitude, frequency, return interval and intensity. Furthermore, kiln charcoal analysis can identify the consequences of such anthropogenic disturbance on the forest ecological trajectories due to changes in forest composition and structure. Recent studies have focused on the effects of historical charcoal production on the functioning of forest ecosystems. For example, it has been shown that the presence of a charcoal-rich soil layer changes the geochemical properties of soils, from the past up to the present day (Hardy et al. 2017; Hirsch et al. 2017; Bonhage et al. 2020). Such long-lasting changes in soil properties, as well as changes in forest composition and structure, are the legacy effects of historical charcoal production on the current state of forest ecosystems. Knowledge about these legacies is crucial to gaining insights that will help guide the future use of ligneous resources, such as an energy transition that could lead to an increase in wood use for biomass energy supply.

9.6. References

Blondel, F., Cabanis, M., Girardclos, O., Grenouillet-Paradis, S. (2018). Impact of carbonization on growth rings: Dating by dendrochronology experiments on oak charcoals collected from archaeological sites. *Quaternary International*, 463, 268–281.

Bonhage, A., Hirsch, F., Schneider, A., Raab, A., Raab, T., Donovan, S. (2020). Long term anthropogenic enrichment of soil organic matter stocks in forest soils – Detecting a legacy of historical charcoal production. *Forest Ecology and Management*, 459, 117814.

Bronk-Ramsey, C. (2016). Methods for summarizing radiocarbon datasets. *Radiocarbon*, 59, 1809–1833.

Carrari, E., Ampoorter, E., Bottalico, F., Chirici, G., Coppi, A., Travaglini, D., Verheyen, K., Selvi, F. (2017). The old charcoal kiln sites in central Italian forest landscapes. *Quaternary International*, 458, 214–223.

Deforce, K. and Haneca, K. (2015). Tree-ring analysis of archaeological charcoal as a tool to identify past woodland management: The case from a 14th century site from Oudenaarde (Belgium). *Quaternary International*, 366, 70–80.

Deforce, K., Boeren, I., Adriaenssens, S., Bastiaens, J., De Keersmaeker, L., Haneca, K., Tys, D., Vandekerkhove, K. (2013). Selective woodland exploitation for charcoal production. A detailed analysis of charcoal kiln remains (ca. 1300–1900 AD) from Zoersel (northern Belgium). *Journal of Archaeological Science*, 40, 681–689.

Deforce, K., Groenewoudt, B., Haneca, K. (2021). 2500 years of charcoal production in the Low Countries: The chronology and typology of charcoal kilns and their relation with early iron production. *Quaternary International*, 593–594, 295–305.

Dufraisse, A. (2006). Charcoal anatomy potential, wood diameter and radial growth. In *Charcoal Analysis: New Analytical Tools and Methods for Archaeology*, Dufraisse, A. (ed.). BAR International Series 1483, Archaeopress Ltd, London.

Dufraisse, A., Coubray, S., Girardclos, O., Nocus, N., Lemoine, M., Dupouey, J.-L., Marguerie, D. (2018). Anthraco-typology as a key approach to past firewood exploitation and woodland management reconstructions. Dendrological reference dataset modelling with dendro-anthracological tools. *Quaternary International*, 463, 232–249.

Dufraisse, A., Bardin, J., Picornell-Gelaber, L., Coubray, S., Garcia-Martinez, M.-S., Lemoine, M., Moreiras, S.V. (2020). Pith location tool and wood diameter estimation: Validity and limits tested on seven taxa to approach the length of the missing radius on archaeological wood and charcoal pieces. *Journal of Archaeological Science Reports*, 29, 102166.

Dupin, A. (2018). Caractérisation du charbonnage moderne et contemporain bisontin (Franche-Comté, France) et de son impact sur les peuplements forestiers : le cas de la forêt de Chailluz. PhD Thesis, Université de Bourgogne Franche-Comté, Besançon.

Dupin, A., Girardclos, O., Fruchart, C., Laplaige, C., Nuninger, L., Dufraisse, A., Gauthier E. (2017). Anthracology of charcoal kilns in the forest of Chailluz (France) as a tool to understand Franche-Comte forestry from the mid-15th to the early 20th century AD. *Quaternary International*, 458, 200–213.

Gebhardt, A. (2007). Impact of charcoal production activities on soil profiles: The micromorphological point of view. *ArcheoSciences*, 31, 127–136.

Gocel-Chalté, D., Guerold, F., Knapp, H., Robin, V. (2020). Anthracological analyses of charcoal production sites at a high spatial resolution: The role of topographical parameters in historical tree taxa distribution in Northern Vosges Mountains (France). *Vegetation History and Archaeobotany*, 29, 641–655.

Grenouillet-Paradis, S. (2012). Etudier les forêts métallurgiques : analyses dendro-anthracologiques et approches géohistoriques. L'exemple des forêts du mont Lozère et du Périgord-Limousin. PhD Thesis, Université de Limoges, Limoges.

Hardy, B., Leifeld, J., Knicker, H., Dufey, J.E., Deforce, K., Cornélise, J.-T. (2017). Long-term changes of chemical properties of preindustrial charcoal particles aged in forest and agricultural temperate soil. *Organic Geochemistry*, 107, 33–45.

Hirsch, F., Raab, T., Ouimet, W., Dethier, D., Schneider, A., Raab, A. (2017). Soils on historic charcoal hearths: Terminology and chemical properties. *Soil Science Society of America Journal*, 81(6), 1427–1435.

Karimi Moayed, D., Vandenberghe, D.A.G., Deforce, K., Bastiaens, J., Ghyselbrecht, E., Debeer, A.-E., De Smedt, P., De Clercq, W., De Grave, J. (2020). Bypassing the Suess-effect: Age determination of charcoal kiln remains using OSL dating. *Journal of Archaeological Science*, 120, 105176.

Knapp, H., Robin, V., Kirleis, W., Nelle, O. (2013). Woodland history in the upper Harz mountains revealed by kiln site, soil sediment and peat charcoal analyses. *Quaternary International*, 289, 88–100.

Larsen, A., Robin, V., Heckmann, T., Fülling, A., Larsen, J.R., Bork, H-R. (2016). The influence of historic land-use changes on hillslope erosion and sediment redistribution. *The Holocene*, 26, 1248–1261.

Ludemann, T. (2010). Past fuel wood exploitation and natural forest vegetation in the Black Forest, the Vosges and neighbouring regions in western central Europe. *Palaeogeography, Palaeoclimatology, Palaeoecology*, 291, 154–165.

Ludemann, T. (2012). Airborne laser scanning of historical wood charcoal production sites – A new tool of kiln site anthracology at the landscape level. *Saguntum: Papeles del Laboratorio de Arqueología de Valencia*, 13, 247–252.

Ludemann, T. and Nelle, O. (2002). *Die Wälder am Schauinsland und ihre Nutzung durch Bergbau und Köhlerei*. Forstliche Versuchs-und Forschungsanstalt Baden-Württemberg, Abteilung Botanik und Standortskunde, Freiburg.

Ludemann, T. and Nelle, O. (2017). Anthracology: Local to global significance of charcoal science. *Quaternary International*, 457, 1–7.

Máliš, F., Bobek, P., Hédl, R., Chudomelová, M., Petřík, P., Ujházy, K., Ujházyová, M., Kopecký, M. (2020). Historical charcoal burning and coppicing suppressed beech and increased forest vegetation heterogeneity. *Journal of Vegetation Science*, 32, e12923.

Mayoral, A., Toumazet, J.-P., Simon, F.-X., Vautier, F., Peiry, J.-L. (2017). The highest gradient model: A new method for analytical assessment of the efficiency of LiDAR-derived visualization techniques for landform detection and mapping. *Remote Sensing*, 9(2), 120.

Nelle, O. (2003). Woodland history of the last 500 years revealed by anthracological studies of charcoal kiln sites in the Bavarian Forest, Germany. *Phytocoenologia*, 33(4), 667–682.

Pain, S. (2017). Power through the ages. *Nature*, 551, 134–137.

Puech, C., Durrieu, S., Bailly, J.-S. (2012). Aireborne LIDAR for natural environments: Research and applications in France. *Revue française de photogrammétrie et de télédétection*, 200, 54–68.

Schneider, A., Takla, M., Nicolay, A., Raab, A. (2015). A template-matching approach combining morphometric variables for automated mapping of charcoal kiln sites. *Archaeological Prospection*, 22, 45–62.

Schneider, A., Bonhage, A., Raab, A., Hirsch, F., Raab, T. (2020). Large-scale mapping of anthropogenic relief features – Legacies of past forest use in two historical charcoal production areas in Germany. *Geoarchaeology*, 35(4), 545–561.

Trier, Ø.D., Salberg, A.-B., Holger Pilø, L. (2016). Semi-automatic mapping of charcoal kilns from airborne laser scanning data using deep learning. In *Oceans of Data*, Matsumoto, M. and Uleberg, E. (eds). Archeopress Publishing LTD, Oxford.

Walker, M. (2005). *Quaternary Dating Methods*. John Wiley & Sons, Chichester.

Žutautas, V. (2017). Charcoal kiln detection from LiDAR-derived digital elevation models combining morphometric classification and image processing techniques. Master thesis, University of Gävle, Gävle.

10

Ancient Trees and Botanical Indicators as Evidence for Change and Continuity in Landscape Evolution

Ian D. ROTHERHAM
Sheffield Hallam University, UK

10.1. Introduction

There is considerable interest in the idea of either "ancient" woods or "old-growth" forest (e.g. Rackham 1976, 1980, 1986, 2007; Peterken 1981, 1996) and in relating this to contemporary management. The conceptual basis of this approach to treed landscapes emerged in the United Kingdom in the 1970s through discussions between Cambridge University scholar Oliver Rackham and government agency woodland advisor George Peterken (Peterken pers. comm. and 2018). These early discussions were a first attempt to rationalize the place of woodlands and "woods" in the historic landscape. Additionally, evidence emerged that these landscapes with long timelines were "ancient" (defined by Rackham as pre-1700 AD or Peterken as pre-1600 AD), but not "primeval" wildwood. These were anciently treed countryside with long histories of human usage and management but ancient woods were not de facto "wildwood" (Rotherham 2007). With wider awareness of the importance of ancient woodlands, the misconception that these equated to wildwood grew. Indeed, many approaches to wooded landscapes and their management were fundamentally flawed by lack of understanding of the historic context of sites and their ecologies (Rotherham 2011). Research in the United Kingdom and across Europe addressed these issues to

Historical Ecology,
coordinated by Guillaume DECOCQ.

provide a robust interrogation of forest and woodland dynamics and better inform contemporary conservation management (Çolak et al. 2018). This chapter presents a brief overview of evidence for these assertions and summarizes a conceptual model into which historical and ecological information can be placed. This allows critical assessment of woodland antiquity and ecological continuity.

10.2. What is ancient woodland? Questions of woods versus old-growth forest, and of continuity versus antiquity

Taking "ancient woodland" in England as defined in England, for example, it is an area wooded continuously since at least 1600 AD, though some definitions give 1700 AD. The choice of this date is explained in Peterken (1977, 2018). Definitions are of course essential but inherently problematic. "Continuously wooded" in this sense does not imply continuous physical cover of trees and shrubs since open land (temporary and permanent) is an important part of a woodland landscape. Furthermore, in all ancient woods, trees and shrubs have been periodically felled and if an area remained woodland, through either replanting or regrowth of, for example, coppiced areas, it is still ancient.

Indeed, and potentially confusing (see Rotherham 2007) if a working wood has been cut-over many times the ancient woodland may have few old standard trees, but may have very old (frequently unrecognized) coppice stools. Guidance in the United Kingdom suggests that if over the last four centuries, woodland has experienced a long period when the land was effectively open (e.g. grassland, heath, moor or arable), then it is "recent" woodland. Such a site may have high nature conservation value, but is not ancient woodland. However, reinterpretation of site-based studies and criteria from an ecological historical viewpoint has provided a more nuanced interpretation of these landscapes (Rotherham 2017) that has implications for the Vera (2000) vision of the European landscape. Multidisciplinary studies have shown how ancient woods have their origins in the enclosure of land from the wider countryside of wood-pastures during the early medieval period. In England, this was associated with the imposition of Norman feudal approaches following the Norman Conquest (1066) and documented in the Domesday account (1086). Additionally, it is apparent that significant areas of unenclosed wood-pastures remain unrecognized in the modern countryside and, through abandonment or degradation, have been largely wrongly assigned by contemporary ecologists.

10.3. The value of ancient woods

Ancient woodlands are of very high ecological and landscape importance and are also hugely significant for their archaeology, historical significance and heritage (see Rotherham et al. 2008). Numerous rare and often threatened species are associated with ancient woods (Peterken 2000), and relatively undisturbed sites may contain important features of historical, archaeological and landscape significance.

Woodlands and their trees form one of the most valuable habitats for wildlife across Britain and Europe (Box 10.1). In lowland situations, an ideal scenario for nature conservation would be mature, mixed woodland including ancient and veteran trees, dead branches and timber lying on the ground, open rides, clearings and glades scattered throughout the wood, with diverse ground vegetation and recognizable field layers of shrubs with no single component dominating. This combination of habitats within the woodland macro-habitat offers the greatest range of conditions to attract the widest variety of wildlife.

1) Exceptionally rich in wildlife with rare species and habitats.

2) Provide high-quality renewable resources of hardwood timber and other woodland products.

3) May hold surviving descendants and features from natural "forests".

4) Reservoirs from which wildlife can spread into newly created woodlands.

5) Integral components of historic landscapes.

6) May have historic features little altered by modern cultivation or macro-disturbance.

7) Contribute to sense of place and imagination for a locale.

8) Importance as features in the landscape.

Box 10.1. *The conservation value of woods and anciently treed landscapes*

10.4. Methodology

This chapter considers (i) methodological issues – approaches to survey, information gathering and analysis (field survey, citizen science, GIS manipulation, LiDAR analysis), (ii) review of literature (indicators, colonization rates, tree forms and aging), (iii) emerging concepts (shadow woods and indicators) and (iv) synthesis of on-going research from case study examples.

10.4.1. *Evidencing ancient woodlands and the use of indicators*

Identifying ancient woodlands and confirming their status is important with countryside under pressures from urban and agricultural or forestry development.

Established practice to identify ancient woodland mixes map-based and field-evidence such as indicators. The latter are species, usually vascular plants, more commonly found in ancient woods than in recent sites. This is evidenced from regional site/species lists. The occurrence of suites of indicator species is used as evidence for woods being ancient, with interpretation interrogated at local and regional levels. No individual plant species is a perfect ancient woodland indicator, and in particular, the degree of association of species with ancient woodland varies across the countryside with geology, topography, climate and land-use factors.

The variety of plants and trees within woodlands depends on soil type and condition and also on sunlight reaching the woodland floor. So, for example, heavily shaded woodland (such as coniferous monocultures and beech-dominated woods) is not conducive to a diverse ground flora. Without diverse flowering plants, invertebrate variety is reduced and populations of small mammals and of birds may be restricted. Beech trees come into leaf before oaks and also cast a dense shade so there is hardly any ground layer under their canopy. Generally, mixed composition and age of woodland are necessary for high nature conservation value (Peterken 1981). This structure includes clearings and glades formed by tree fall or loss of major limbs from large mature trees. When applied systematically over a long period, traditional woodland management such as coppicing creates varied light conditions. Recently coppiced compartments are opened to sunlight, whereas later in the cycle, under re-grown springwood conditions are much darker. The impact of management history has major implications for the likelihood of indicator plant presence and visibility. It also affects invertebrates and other indicators both indirectly through vegetation and directly through microclimatic conditions.

Coppicing involves cutting and removing "wood" from the woodland environment. In the past, this included "tidying up" coppice stools and in the working wood, having a "clean" housekeeping approach with clearance of dead fallen limbs and "ramel" left after extraction. Saproxylic invertebrates (associated with dead wood) are adversely affected by such disturbance and removal of habitat. The stage of dead wood decay is important with some invertebrates associated with recent dead wood and others with well-rotted material in dead standing trees. Ground-layer flowering plants are important nectar and pollen sources as are flowering trees and shrubs. Mature hawthorn and wild fruit trees such as crab apple and wild pear, for instance, are important food sources for adult deadwood beetles, craneflies and hoverflies whose larvae feed on ancient trees. Management for traditional and industrial products such as charcoal and "white-coal" (kiln-dried wood used for metal-smelting) also involved the stripping of vegetation and soil from the woodland floor to cover charcoal clamps and white-coal kilns. This resulted in intensively managed early industrial woods with very limited diversity of flowering plants, and soils with depauperate upper horizons and sometimes replaced entirely by a shallow layer of pure charcoal dust. Botanical indicators can inform judgments of both woodland continuity (through presence) and past management (through absence).

Anemone nemorosa (wood anemone)
Paris quadrifolia (herb Paris)
Lamium galeobdolon (yellow archangel)
Mercurialis perennis (dog's mercury)
Circaea lutetiana (enchanter's nightshade)
Conopodium majus (pignut)
Veronica montana (wood speedwell)
Lysmachia nemorum (yellow pimpernel)
Allium ursinum (wild garlic)
Luzula pilosa (hairy woodrush)
Luzula sylvatica (greater woodrush)
Oxalis acetosella (wood sorrel)
Hyacinthoides non-scriptus (bluebell)
Narcissus pseudonarcissus (wild daffodil)
Melampyrum pratense (common cow-wheat)
Corydalis claviculata (climbing corydalis)
Chrysosplenium oppositifolium (opposite-leaved golden-saxifrage)
Lonicera periclymenum (honeysuckle)
Millium effusum (wood millet)
Melica uniflora (wood melick)
Melica nutans (Mountain melick
Valeriana officinalis (common valerian)
Orchis mascula (early purple orchid)
Primula vulgaris (primrose)
Carex remota (remote sedge)
Carex sylvatica (wood sedge)
Carex pendula (pendulous sedge
Galium odoratum (woodruff)
Holcus mollis (Creeping soft-grass)
Deschampsia flexuosa (wavy hair-grass)
Festuca gigantea (giant fescue)
Geum urbanum (wood avens)
Ilex aquifolium (holly)
Corylus avellana (hazel)
Prunus padus (bird cherry)

Table 10.1. *Exemplar botanical indicators of ancient woodland sites – based on the English Peak District and Rotherham et al. (2008), Rotherham (2011, 2017). The use of indicator lists must be regionally validated*

It is important to recognize the diversity of woodland types and their origins that are broadly: medieval and industrial coppice, park and pasture woods, ancient forest, wooded common, linear remnant and fragments (Rotherham 2007, 2017). Evidence ancient status, and particularly the use of botanical indicators, generally relates to former coppice woods (Rotherham 2011, 2013). Ongoing studies over 30 years, long-term action research with expert stakeholders and detailed regional audits allowed a review of evidence. The two-year project on the Woodland Heritage Manual (Rotherham et al. 2008) evolved into a series of expert stakeholder workshops examining and reviewing woodland indicators, woodland inventories and associated landscape issues. This research reviewed the use and interpretation of indicator species, of inventory lists and the advocacy of holistic evidence-based evaluation.

Ancient woods are initially identified on map bases. In England and Wales, the main cartographic sources used to identify potential ancient woodland sites were 1:25,000 maps from the 1920s and 1930s, the first edition one-inch Ordnance Survey maps (generally mid-1800s), aerial photographs and, if available, survey reports too. In the current work, the use of LiDAR imagery and GIS computer mapping helps enormously. Ground-truthed where possible, most assessments of ancient woodland status were originally from research in the late 1970s and early 1980s, and essentially intuitive lists generated by known local experts and inspection of First Edition Ordnance Survey Maps (dating around 1830 to 1840). If available, earlier documentation supported assessments and most sites had ecological surveys over the following 10 to 15 years. These varied in merit depending on who the surveyors were and why they were done. Until the 2008 Woodland Heritage Manual (Rotherham et al. 2008), few sites had evaluations of additional factors like soils, woodland archaeology and heritage, or historic interest.

From the review, it was clear that the broader context of historic sources and archaeological evidence was often overlooked and even approaches such as interpretation of map-based place-names, for example, were not widely used.

10.4.2. *Tree form and growth as evidence of antiquity and continuity*

Furthermore, ancient woods which lacked large, old-looking, standard trees were frequently overlooked. Given that former medieval coppice woods, because of their history of management inherently lack veteran standard trees, this is a serious flaw. Additionally, there is a lack of accepted, robust approaches to aging old coppice stools, and reluctance by many organizations to recognize these as ancient or veteran. Yet the presence of verifiable ancient coppice stools can provide confirmation of woodland status going back centuries. Furthermore, such trees are evidence of both the antiquity of a site and of former management. Other trees which are clonal or naturally self-coppicing (such as *Ilex aquifolium*, *Alnus glutinosa* or *Tilia cordata*) may provide further information on the landscape timeline.

Figure 10.1. *Ancient oak coppice stool computer model reproduced with permission from Vrška et al. (2016). For a color version of this figure, see www.iste.co.uk/decocq/ecology.zip*

Recent work in the Czech Republic (Vrška et al. (2016) suggests oak coppice (*Quercus robur*) stools up to 825 ± 145 years in age (Figure 10.1), and this has implications for dating both coppice woods and shadow wood "medusoid" oaks in relict treescapes (Rotherham 2017). Along with the dating of small-leaved lime coppices at well over a thousand years in, for example, Cumbria in England (Donald Pigott pers. comm.), and the recent recognition of ancient but small veteran specimens of hawthorn (*Crataegus monogyna*) and rowan (*Sorbus aucuparia*) in shadow woods (Rotherham 2017), this work provides evidence of treescape antiquity, continuity and persistence.

10.4.3. *The importance of ancient and veteran trees in woodland*

The presence of ancient and veteran trees in woodland adds a highly valuable habitat component for many uncommon and rare species, particularly invertebrates (Box 10.2).

Ancient: A tree that is very old, in the declining stages of life, and usually relatively large in girth in relation to other trees of its species. This depends on how and where it has grown. Ancient trees are not always particularly tall but stand out visually as being very special. Worked trees (i.e. those formerly managed to produce particular crops and products) can be difficult either to recognize or to date. This applies especially to former coppice trees or to some smaller ancient trees.

Veteran: A tree usually in a mature stage of life and with important wildlife and habitat features including: hollowing or associated decay, fungi, holes, wounds and large dead branches. These include old trees as well as younger middle-aged trees with premature aging characteristics. These processes with species and with levels of environmental stress, for example, upland or lowland status.

Box 10.2. *"Ancient" and "veteran" trees (Rotherham et al. 2008)*

10.4.4. *Soils and sediments*

Evidence for woodland history can be found in soils, sediments and other deposits, and yet most ecologists overlook these. Indeed, until relatively recently, most paleoecologists and paleobotanists dismissed wet sediment deposits in ancient woods as being too fragmented and disturbed to be analyzed. However, current work is showing that timelines of pollen and insect remains can be retrieved from sites to help re-construct vegetation and faunal histories over many centuries (Helen Shaw pers. comm.). Furthermore, such evidence adds to work on archaeological remains which also provide information on land-use and disturbance. The replacement of woodland soil by layers of charcoal provides insight into intensive usage for woodland industries (Rotherham 2007). The earthworks within a wooded landscape can give remarkable insight into human usage from modern times back to prehistory (Rotherham 2007). These may be evidence of long periods of non-woodland status with, for example, Roman-British or Dark Ages cultivation features, and periods of both "coppice wood" management with woodbanks and deer park usage with hunting structures.

In order to identify and map features such as earthworks, the application of LiDAR (Light Detection and Ranging), aerial photography, GPS (Global Positioning System) and drone flyover has revolutionized what is possible. With sites identified from images they can be located and mapped on the ground in order to be ground-truthed. From these information sources, computer-generated maps GIS (Geographic Information System) and terrain models can also be created. These technological tools can be combined with ecological and historical (archive) information to provide a robust and refined assessment of a treescape site.

Furthermore, data gathered on, for example, botanical indicator species distributions can be used to create GIS "heat-maps" and thus to reconstruct possible past landscapes from the modern data.

10.5. An emerging woodland paradigm

What we define as "woodland" and as "ancient" is not as simple as might be assumed. Most sites recognized through both indicators and historical sources are medieval coppice woods enclosed from former wood-pasture or wooded common. Lack of recognition means that many sites with different timelines and histories may be excluded from ancient status, and the consequences for conservation and management may be serious. There are two main issues arising from the work so far. The first is the need for robust indicators and second, for these to be placed in a broader framework for integration and interrogation.

However, with the exceptions of obvious pollarded trees, for example, "worked" trees, "modified" trees and naturally contorted trees (such as by animal grazing and/or extreme weather) have tended to be overlooked. Coppice trees such as ancient limes, alders and willows have been mostly neglected except for the pioneering work of Donald Pigott (pers. comm.). Work on some species such as sweet chestnut is currently in process, but there is a tendency to overlook smaller species and peripheral habitats such as "shadow woods" (Rotherham 2017). Upland sites and coppices are often ignored, which is problematic. Whilst understanding and aging these specimens may be difficult, such trees are intimately tied to human countryside exploitation and to extreme weather and climate. Many of these specimens are older than "standard" trees, which are more generally recognized as significant.

10.6. A simple new conceptual framework

The approaches described can be incorporated into a holistic model to bring a multidisciplinary evidence-based approach to ancient woodland assessment (Rotherham 2011). The basic steps are indicated in Figure 10.2.

The enhanced understanding of woodland origins, antiquity, continuity and change can then be placed into a simple conceptual framework in relation to the ideas postulated by Vera (2000) of the European open savanna primeval landscape (Figure 10.3).

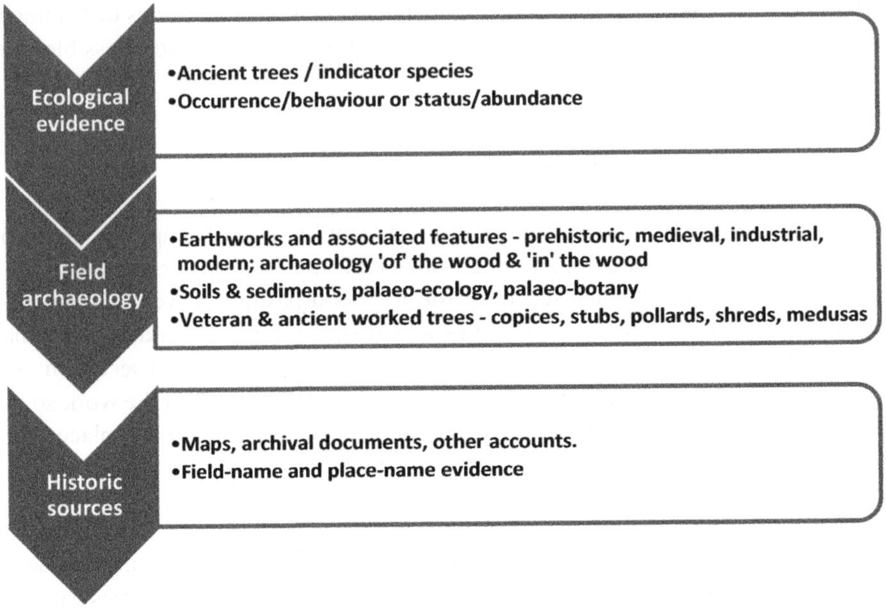

Figure 10.2. *Evidencing ancient woodland. For a color version of this figure, see www.iste.co.uk/decocq/ecology.zip*

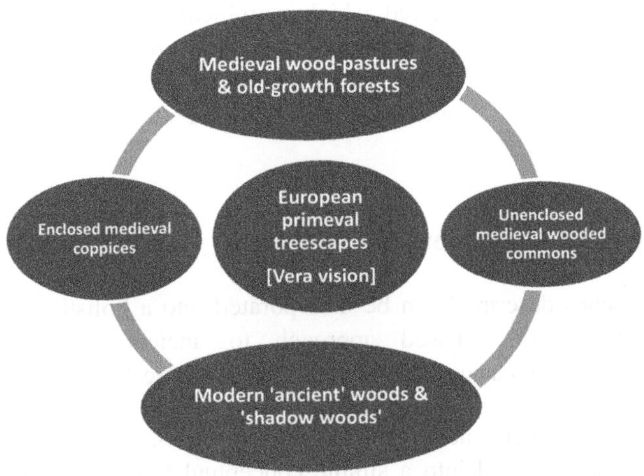

Figure 10.3. *Simple conceptual framework for woodland origins, continuity and landscape change in forests. An evidence-based model derived from Rotherham (2011, 2017). For a color version of this figure, see www.iste.co.uk/decocq/ecology.zip*

10.7. Conclusion

Recognition of "ancient woodland" diversity has implications for future protection, conservation and management. However, the concept evolved piecemeal to leave sites vulnerable to damage or destruction through inappropriate development. A more holistic evidence-based approach drawing on assessments of botanical indicators, veteran trees, soils, archaeology and historical records is more robust. The conceptual framework is supported through a mixed-methods approach to information gathering and evaluation. Importantly, through the conceptual framework and the emergence of the "shadow woods" concept, we are able to place "woods" in their historic timeline in relation to the wood-pasture and old-growth forest origins of European treescapes (Vera 2000). Furthermore, the evidence for both continuity and change in these landscapes is significant in developing and applying realistic future policies to treescape management.

10.8. References

Çolak, A.H., Kirca, S., Rotherham, I.D. (eds) (2018). *Ancient Woodlands and Trees: A Guide for Landscape Planners and Forest Managers*. Turkish Academy of Sciences, Ankara and IUFRO, Vienna.

Peterken, G.F. (1974). A method for assessing woodland flora for conservation using indicator species. *Biol. Conserv.*, 6, 239–245.

Peterken, G.F. (1977). Habitat conservation priorities in British and European woodlands. *Biol. Conserv.*, 11, 223–236.

Peterken, G.F. (1981). *Woodland Conservation and Management*. Chapman and Hall, London.

Peterken, G.F. (1996). *Natural Woodland: Ecology and Conservation in Northern Temperate Regions*. Cambridge University Press, Cambridge.

Peterken, G.F. (2000). Identifying ancient woodland using vascular plant indicators. *Br. Wildl.*, 11, 153–158.

Peterken, G.F. (2018). Ancient woodland in concept and practice. In *Ancient Woodlands and Trees: A Guide for Landscape Planners and Forest Managers*, Çolak, A.H., Kirca, S., Rotherham, I.D. (eds). Turkish Academy of Sciences, Ankara and IUFRO, Vienna.

Rackham, O. (1976). *Trees and Woodland in the British Landscape*. J.M. Dent & Sons Ltd, London.

Rackham, O. (1980). *Ancient Woodland: Its History, Vegetation and Uses in England*. Edward Arnold, London.

Rackham, O. (1986). *The History of the Countryside*. Dent, London.

Rackham, O. (2007). *Woodlands*. Collins New Naturalist, London.

Rotherham, I.D. (2007). The implications of perceptions and cultural knowledge loss for the management of wooded landscapes: A UK case-study. *For. Ecol. Manage.*, 249, 100–115.

Rotherham, I.D. (2011). A landscape history approach to the assessment of ancient woodlands. In *Woodlands: Ecology, Management and Conservation*, Wallace, E.B. (ed.). Nova Science Publishers, Inc., Hauppauge, New York.

Rotherham, I.D. (2013). *Ancient Woodland: History, Industry and Crafts*. Shire Publications, Oxford.

Rotherham, I.D. (2017). *Shadow Woods: A Search for Lost Landscapes*. Wildtrack Publishing, Sheffield.

Rotherham, I.D., Jones, M., Smith, L., Handley, C. (eds) (2008). *The Woodland Heritage Manual: A Guide to Investigating Wooded Landscapes*. Wildtrack Publishing, Sheffield.

Vera, F.W.M. (2000). *Grazing Ecology and Forest History*. CABI Publishing, Oxon.

Vrška, T., Janík, D., Pálková, M., Adam, D., Trochta, J. (2016). Below- and above-ground biomass, structure and patterns in ancient lowland coppices. *iForest*, 10, 23–31.

11

Towards a Methodological Framework for Investigating the Hidden History of Woodland Covers

Damien MARAGE[1], Catherine FRUCHART[1], Isabelle JOUFFROY-BAPICOT[1], Olivier GIRARDCLOS[1], Vincent BALLAND[2]

[1] *Bourgogne-Franche-Comté University, Besançon, France*

[2] *Bourgogne-Franche-Comté University, Dijon, France*

11.1. Why talk about hidden history when studying forest vegetation?

In the wake of the craze for environmental studies in history and historical ecology, forest has become an interdisciplinary field of studies. Results in environmental sciences are confronted with historical and archaeological data. Nowadays, if the ecology of present forest vegetation is well documented, historical and archaeological studies remain rare, mostly because field surveys are hampered by the presence of vegetation and also because studying historical data is time-consuming: archives are plethoric and scattered among many private and public funds. Thus, our perspective on forests is, in a way, still "hidden". This contribution is a short review of the major disciplines and approach that we propose to cover for the studies of historical ecology of forests. It mentions works that have already been carried out with such approaches in France. We finally suggest a methodological framework that could unravel the "secrets" of European forests vegetation history over the Holocene period.

Historical Ecology,
coordinated by Guillaume DECOCQ.

11.2. From recent forests: a synecological point of view

Reconstructing past forest dynamics presupposes an understanding of the present forest dynamics. The study of natural forest ecosystems is a key step to achieve this reconstruction.

Historically, the complex unraveling of ecological interactions was developed in the classic paper on pattern and process in forest communities by Watt (1947). One conspicuous introduction of these concepts was the review made by Whittaker (1953) which used the Watt pattern-and-process paradigm to redefine the "climax concept", and furthermore to link life trait history and biogeography.

Under unconstraint conditions, the light available under the tree canopy is reduced and the presence of the tree considerably modifies the microclimate. These conditions determine which tree species can survive underneath the canopy. The establishment of a new dominant canopy represents the closure of the death/birth/death cycle that can be thought of as the typical small-scale disturbance of a forest. According to Van Steenis (1956), tree species should be classified into three functional groups: (i) pioneers, (ii) post-pioneers and (iii) dryads. One characteristic of the mortality of trees in the canopy, and the associated openings ("gap formation"), is the size of the created gaps. Species attributes are important to differentiate the gap-size-related regeneration success of various forest ecosystems.

The persistent seed bank is a dominant driver in community assemblages, especially in the earliest stages of secondary succession next to vegetation extant or seed rain, for example. The seed bank remains important long after the canopy has closed, standing as a declining input along canopy's development and maturation. Seeds may be recruited from the persistent seed bank in any woodland with recurrent short-interval disturbances as coppiced stands, with forest canopy gaps or small gaps in the herbaceous layer. The latter increasingly contributes to viewing seed banks as a functioning part of non-successional communities in temperate deciduous forests (Plue et al. 2010). For ecological processes, three main categories in forest historical ecology have to be considered: (i) tree functional groups, (ii) forest-gap dynamics and (iii) soil seed bank dynamics.

11.3. From the walls: ancient documents and maps

Textual archives and historic maps are essential to reconstruct forest landscape history and to understand how history legacies may influence the current forest vegetation. In France, historians have used a regressive analysis to study beyond the French historic "Eaux et Forêts" reformation, in 1669. Ancient archives from the pre-industrial period are difficult to analyze. They require special skills in paleography and in linguistics for their deciphering and good understanding (Figure 11.1). The history of French forests is well documented, and we shall examine the major studies and sources.

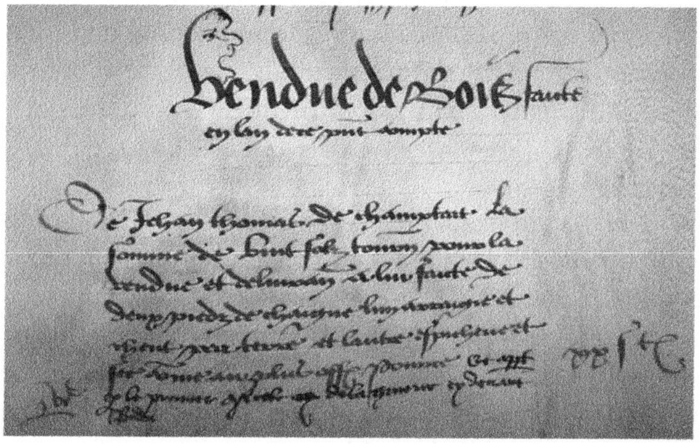

Figure 11.1. *1474–1475, Accounting of the châtellenie de Glenne (Saône-et-Loire, France), Gruerie section, sale of wood, Archives départementales de la Côte d'Or B 4879-2, fol. 28 r. For a color version of this figure, see www.iste.co.uk/decocq/ecology.zip*

"Modern" maps (18–19th century) provide substantial data covering the whole of France. The "Etat-Major" maps, produced between 1820 and 1866 on the entire metropolitan area, are reference maps that can be used by ecologists to identify ancient forests, as opposed to "recent" ones (Bergès and Dupouey 2017).

The "Napoleonian cadaster" was realized during the first half of the 19th century in order to inform on tax and property data for all the French communes. The maps are composed of thousands of separate sheets giving an accurate planimetry of land parceling and indicating toponyms. Its matrix and tables detail land-use partly composed of private and public wood covers. They also provide the names of owner, plots surfaces, function and land-use also including information on the forest nature and regime. The archives of the "Eaux et Forêts" also provide interesting data about public forests management plans of the 19th to 20th century (Corvol 1999). During the French Revolution, at the end of the 18th century, documents with accurate information were produced through the sale of confiscated properties, previously owned by ecclesiastic institutions and noble people.

The Cassini map was drawn up during the second half of the 18th century over the entire kingdom of the time. Its main purpose was to represent the most important traveling axes and the various inhabited places of the country. Because of its lack of precision, it is difficult to use it for the study of woodland cover change because forest boundaries were generally roughly represented and the restoration of forest cover was poorly reliable, especially for areas located on reliefs or for woodland of little importance. Apart from this map, especially for the period from the end of the 17th century to the end of the 18th century, the seigneurial and estate woods were generally mapped and represented forest cuttings and gave precise elements on cantonment and on the capacity and layout of the "reserve quarter" and regulated cuts (Rochel 2017).

Additionally, textual archives written between 1661 and 1680 in relation with the French reformation of forests provide precise data on forests condition, forest stands and management (Poublanc 2015). These documents can help reconstructing the past forest management, tree conditions and stands characteristics in royal forests and, sometimes, in religious or secular community woodlands. Seigneurial accounts from serial and annual sources, can help to reconstruct, sometimes over long periods, forest management methods and harvesting. They inform on the frequency of felling and the importance of grazing and are of great help in measuring the anthropogenic activities. In some regions, it is possible to go as far as the end of the Middle Ages like in Burgundy, for example, where we have access to the accounts of the Duchy of Burgundy from the 14th century (Beck 2008); the parchment rolls of the County of Savoy provide information on forest management in county forests since the last quarter of the 13th century.

For areas that were dependent on the kingdom, or for lay or ecclesiastical communities after the 1669 Ordinance, the documentation could be cross-referenced or completed with the minutes of issue of the "Eaux et Forêts" cutting permits. These documents record auction stages and methods of cutting and emptying and control visits or resetting that were carried out, followed by logs re-examination.

The archives for private woods are less frequent. However, public archives departments occasionally keep bundles of titles attached to ecclesiastical or secular lordships. These collections contain various deeds, sometimes dating from the Early Middle Ages. They can help to characterize and evaluate how forestry practices were used by residents. Burrows, documents issued from the 15th century, give many details about woodlands that are unrelated to the seignorial domain, but linked to the peasantry's censive.

Complementary information can be obtained by other sources of documents where social aspects of the woodland landscape are involved. Court and notarial records that settled conflicts often reveal past socio-economic frameworks and attest to methods used to cut down woodland. They give an overview of the professions that revolve around wood like logging contracts, the market of transports, wood purchase and contracts for the processing of wood products.

11.4. From the wood: dendrochronology

The acyclical nature of droughts gives the opportunity to use them as temporal benchmarks and build long chronologies called masters that are useful afterwards for dating many archaeological sites. Regional master curves are based on thousands of oaks (*Quercus robur* and *Q. petraea*), aged 1 to 200 years, discovered both in natural deposits and in archaeological settlements. They collectively cover the entire Holocene period since 10,430 BP. In optimal cases, felling-dates yearly identified are linked to constructions, repairs of houses and/or villages. It *pushed further into the past the horizons of history* to the Neolithic (Petrequin et al. 1998).

Dendrochronology provides variate ecological records. Mathematical tools had made possible the study of the tree response to climate, allowing numerous dendroclimatic reconstructions of temperatures and rainfall variabilities, at an annual resolution (Büntgen et al. 2011). The major result was to highlight an increasing radial growth trend during the 20th century at a global scale and provide climatic feedback. For example, in France, this trend was identified for seven forest species belonging to the entire panel of forests ecosystems (Badeau et al. 1996).

Growth rings also reveal forest stands characteristics. For example, radiocarbon dating up to 500 years tauzin oak stumps (*Q. pyrenaica*), historically coppiced, contrast with sprouts which only presented a few dozens of rings (Salomón et al. 2016). Tree-ring methods have been tested to detect past disturbances, such as canopy disturbance. A study by Altman et al. (2013) established a link between growth characteristics and past forest management, as ring-width profiles responded to recurrence and intensity of thinning.

Dendroecological interpretation of archaeological woods provides records over a larger timescale than with living trees. According to dendrochronological data and historical resources, long-distance transport of valuable timber from production areas to deforested areas was already highly developed during the Middle Ages from countries around the Baltic Sea across western Europe (e.g. Haneca et al. 2005). France remained at the margin of international transport, as timber was exported from mountains to foothills. At least, timbers were floated on the river Moselle, from the upper river Meuse and Yonne basin or from the river Durance to the Rhone corridor.

Finally, dendrochronologists describe the growth of trees exploited in the past in order to characterize the resources needed for constructions. The shape of a tree growth curve could provide valuable information on the levels of competition to which it had been subjected during the different phases of its life, and therefore on the management (or non-management) of the stand where it grew up (e.g. Billamboz 2014). For example, it is established, for several antique and medieval sites (Girardclos and Perrault 2019), that the growth of oaks from the age of 20–30 years was below the growth of current oaks managed in high forest and could be explained by low forests thinning in the past and an unclear distinction between thinning and logging phases.

The case of charcoal remains is interesting within interdisciplinary approaches. Largest pieces give the opportunity to analyze the size of cutting and growth rings. The remains of charcoal from industrial sites, such as charcoal kilns platforms, which can be studied by archaeologists through spatial analysis, tend to indicate that gathering linked to the required volume of wood is very local (Ludemann et al. 2004). The charcoal produced is eventually exported.

In these studies, the ecological information extracted from ancient woods was obtained by comparison with the growth of living trees in known contexts (Dufraisse et al. 2018). In principle, the hypothesis that the biological mechanisms of growth are unchanged between ancient periods and today could be tested. Without looking for current relics of forests from the past, the goal was to compare measurements made on ancient trees with biological models built on living trees. There are benchmarks that make it possible to quantify the density of stands by knowing both the diameter and height growths.

11.5. From the ground: palynology

Pollen analysis has been used for more than a hundred years for the reconstruction of vegetation and the study of dynamics and distribution changes in Quaternary vegetation (Seppä 2007). Many continental sequences allowed to follow forest change from the Late glacial (ca. 17,000 to 11,500 BP), and even more, to the Holocene period (ca. 11,500 BP to present) with tree migrations from refugees in the Mediterranean and the Balkans, in response to warming. Forest recolonization in Europe followed a general trend (Figure 11.2): (i) a pioneer dynamic with pine and hazelnut in the Early Holocene (11,500–8,200 BP), (ii) the establishment of mesothermophilous oak forests during the Climate Optimum of the Mid-Holocene (8,200–4,200 BP) and (iii) the development of oak-beech forest, often accompanied by fir and/or spruce in mountain areas, at the favor of wetter and cooler climate conditions during the Late Holocene (4,200 BP to present). Hornbeam was the last tree species arriving "naturally" in Western Europe about 2,800 years ago. At this time, human influence on vegetation had become prominent since the Neolithic, ca. 7,000 years ago (e.g. Iglesias et al. 2019). Human activity impacts on woodland included forests clearance for the establishment of cropland and pasture, forest exploitation for fuel and timber and proto industry, for example, metalworking. New species importation, such as walnut and chestnut, due to contacts with the Mediterranean world during the Antiquity (Zohary and Hopf 2000), more than 2,000 years ago, also modified woodlands composition.

In recent decades, many local to regional analyses have shed light on woodland evolution at the rhythm of socio-ecosystem construction in Western Europe. These recent palynological studies took place in a framework of multiproxy approaches on cores such as sediments, micro-charcoals, plants macro-remains analyses and, recently, on ancient DNA. Cores dating are increasingly precise with many radiocarbon dates and the use of radioelements on the uppermost part of the cores (^{210}Pb and ^{137}Cs) (Appleby 2001). Another challenging target of recent studies is the understanding of past biodiversity and its long-term dynamics using pollen diversity (e.g. Colombaroli and Tinner 2013).

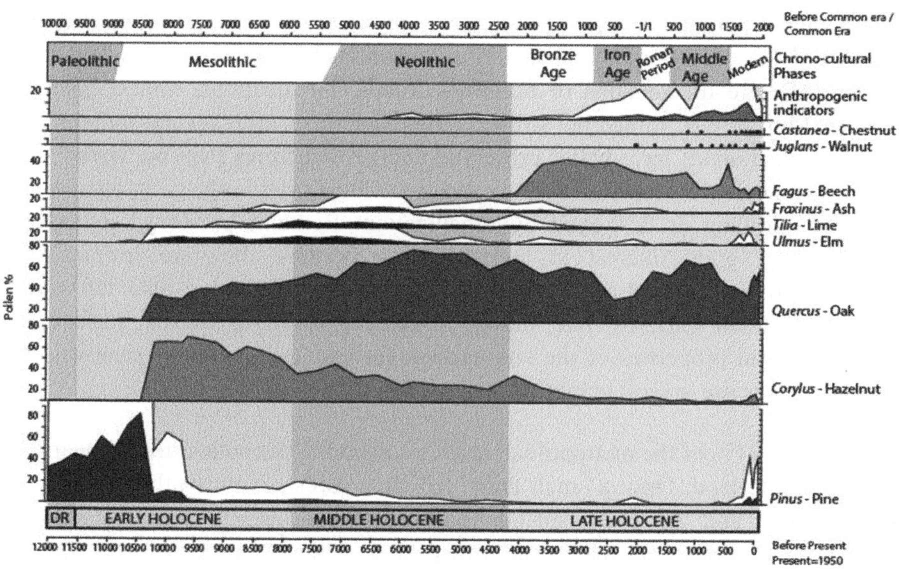

Figure 11.2. *Holocene forest evolution in lowland temperate Europe: the example of Premery sequence (340 m. asl), Burgundy, France. The main tree taxa pollens are expressed here as percentages of total land pollen; anthropogenic indicators curve gathered pollen of crops, weeds and ruderal species (modified from Jouffroy-Bapicot 2014). For a color version of this figure, see www.iste.co.uk/decocq/ecology.zip*

Many studies aim to over cross pollen analysis limits related to pollen identification that rarely reached the species level, and pollen taxa rates that cannot be directly transformed into land cover (e.g. Seppä 2007). The development of models and their application on databases such as the EPD (http://europeanpollendatabase.net/index.php) specifically focus on the development of quantified and spatially explicit approach of the data (e.g. Marquer et al. 2017). At a European scale, Roberts et al. (2018) provided a recent synthesis of these three main different approaches, by transforming fossil data into quantitative past European forests cover over the last 11,000 years.

11.6. From the air: LiDAR

LiDAR is a remote sensing technique providing highly accurate 3D models of physical bodies with mobile laser beams. Archaeologists have used this technique since the early 2000s (Devereux et al. 2005). This data offers an unprecedented opportunity to the detection, studying and characterization of past land-use in current forests. Artificially stripped from plant cover, grounds testify of past anthropogenic activities, still visible at the surface as microreliefs and reveal an amazingly diverse and abundant heritage which is preserved in forests over past centuries and millennia. Traces of anthropogenic activities, when visible as microreliefs detectable with LiDAR, are the remains of buildings, roads, fortifications, mines, quarries, kilns, field limits, etc., ruined but still existing, sometimes after several millennia of abandonment. Forests have better preserved ancient structures than cultivated soils because archaeological remains were erased from landscapes long ago in open areas, eroded by plowing or destroyed by urbanization. Remains preserved in today's forests are of great interest to archaeologists because they are irreplaceable testimonies of past societies. Past anthropogenic activities have also left traces undetectable with LiDAR, like buried structures invisible to the surface, or physicochemical properties of the soil caused, for example, by stone removals for agricultural purpose, or soil manuring.

In France, 31% of the metropolitan area is covered by woodland today, nearly as much as arable land. Only a small part of French forests is currently documented with LiDAR, and an even smaller part has been archaeologically studied. Investigated forests reveal many traces of past activities of various forms like the frequent remains of ancient field systems (Figure 11.3). They testify of agricultural or pastoral activities over the last millennia, often dated from the Roman period (e.g. Georges-Leroy et al. 2012; Fruchart 2014). Our forests also shelter many other forms of remains (e.g. Fruchart 2014; Costa et al. 2020): burial sites and buildings, hillforts, military sites, hundreds of ancient road sections, water management structures, mining remains, quarries, charcoal kilns platforms and limekilns, etc.

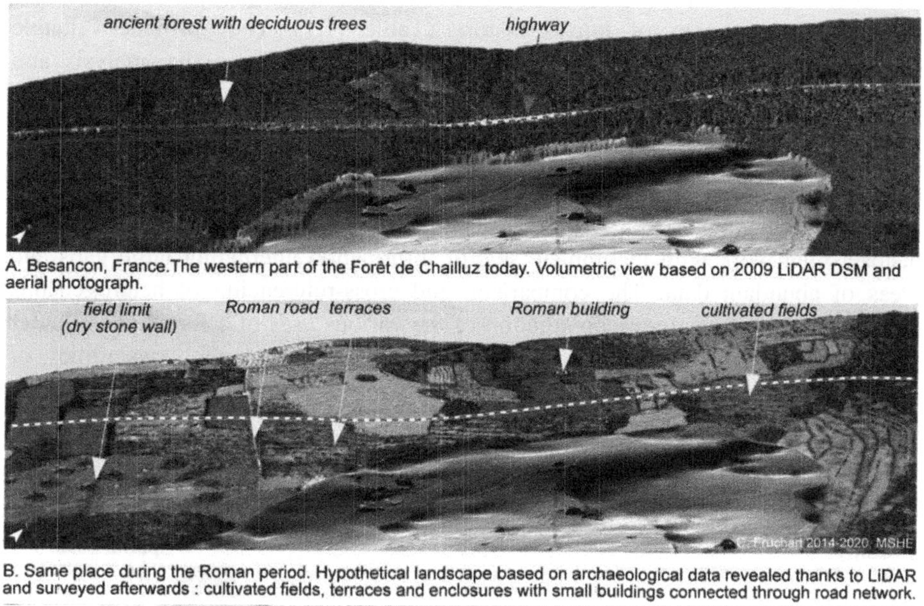

Figure 11.3. *Comparison between past and present landscapes: the Forêt de Chailluz in Besancon, France. (A) Today. (B) Hypothetical rendering for the Roman period. For a color version of this figure, see www.iste.co.uk/decocq/ecology.zip*

Archaeological remains are discovered all over the world in today's forests, covering large areas, with unsuspected extents and sometimes in unforeseen locations (e.g. Opitz and Cowley 2013). This new and abundant information, provided thanks to LiDAR, underlines that we still have little knowledge of the ways ancient societies used their lands – and forests are particularly poorly known from this viewpoint. This field of research needs to be developed through interdisciplinary approaches, to characterize the extents, the patterns and practices associated with past land-use, and to better understand how anthropogenic activities, combined with natural dynamics, contributed at landscapes and environment change over time.

11.7. Discussion

As far as the stability of the two major mechanisms causing ecological niche differentiation over tree life cycle is concerned, i.e. ontogenetic niche shift and environmental changing conditions during the Holocene period, these five disciplines and their toolboxes, when crossed used, allow a better analysis, understanding and interpretation of the effects of anthropogenic activities on patterns and processes of forests history (Table 11.1). They provide valuable complementary information that foster a holistic approach to the analysis and interpretation of evolutionary processes in historical ecology.

Ecological research on woodlands is now widespread all over Europe. Historical archives offer more variable information that strongly depends on the type of available documents as well as on the local or regional context. Specialists of forest history often work with incomplete sets of information alternating large gaps and areas of abundant data. The comparison and cross-referencing of heterogeneous documents allow a better reconstruction of the various uses of a forest and a better visibility of forest developments and dynamics. Likewise, the possibilities of dendrochronology and palynology depend on the nature and context of archaeological sites or sedimentary deposits.

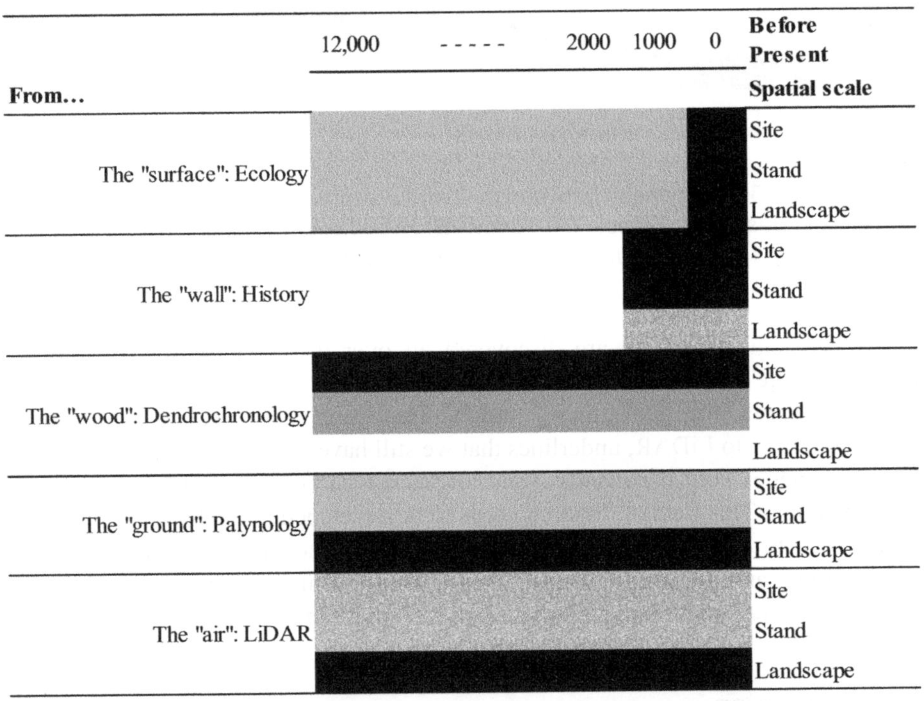

Table 11.1. *Fitting temporal and spatial scale according to disciplines in forest historical ecology. Blank = not suited; gray = weakly suited; hatched = moderately suited; black = highly suited*

Nevertheless, several studies using this type of methods have already been carried out in France in order to answer ecological questions or to focus on historical, archaeological or paleoenvironmental topics. Some major publications emphasize the importance of interdisciplinary research that includes ecology, history, archaeology and paleoenvironmental sciences (Bergès and Dupouey 2017). Some projects that were developed in varied geographical areas and contexts are mentioned in Table 11.2.

Study areas	Disciplines involved	Research topics
Northern France Lowlands (alluvial plains)	Pedology, anthracology, archaeology, history, LiDAR, for example, Feiss et al. (2017)	Understand the dynamics of forest and human relationships in a temperate deciduous forest since the Roman period.
North-eastern France Lorrain plateau (sandstone, limestone) Jura low mountains and Burgundy plateau (karstic limestone)	Ecology, pedology, palynology, dendrochronology, anthracology, archaeology, history, LiDAR, for example, Dupin et al. (2017)	Characterize land-use change since the Roman period and evaluate environmental consequences and legacies to vegetation.
Massif Central Natural Regional Parks (crystalline rocks, limestone)	Ecology, archaeology, history, LiDAR, for example, Renaux and Villemey (2016)	Research on land cover change, ancient forests and associated biodiversity.
South-eastern France Forest of Sainte-Baume, Natural Regional Park of Luberon (limestone)	Ecology, paleoenvironmental approaches, archaeology, history, for example, Abadie et al. (2021)	Evolution of forest management and of oak (*Quercus pubescens*) and beech (*Fagus sylvatica*) presence over several centuries. Land-use legacies in current understory vegetation; analysis of historic archives contribution to the ecological investigation.
South-western France, Pyrenees (crystalline rocks, limestone)	Ecology, forestry, pedology, dendrochronology, anthracology, palynology, archaeology, history, for example, Saulnier et al. (2020)	Multiple interdisciplinary studies since the 1980s, focused on the history and evolution of woodlands, under different headings.

Table 11.2. *French examples of studies involved interdisciplinary in forest historical ecology*

This spatial and quantitative reconstruction is necessary for optimizing the cross-correlation with other data from archaeology and historical archives. The complexity of datasets often demonstrated a continuity of land-uses at the landscape level. Paleoecological insights on the evolution from natural to cultural woodland provide invaluable information for the present and future forest management (Conedera et al. 2017). An increasing number of tools and datasets now available including data mining, increasingly accurate dating for paleoecological data and accessible statistical tools could help achieve the matching of spatial and chronological scales of the different scientific fields.

11.8. References

Abadie, J., Dupouey, J.-L., Salvaudon, A., Gachet, S. (2021). Historical ecology of Mediterranean forests: Land use legacies on current understorey plants differ with time since abandonment and former agricultural use. *J Veg Sci*, 32(1), e12860.

Altman, J., Hédl, R., Szabó, P., Mazůrek, P., Riedl, V., Müllerová, J., Kopecký, M., Doležal, J. (2013). Tree-rings mirror management legacy: Dramatic response of standard oaks to past coppicing in Central Europe. *PLoS One*, 8(2), e55770.

Appleby, P.G. (2001). Chronostratigraphic techniques in recent sediments. In *Tracking Environmental Change Using Lake Sediments*, Last, W.M. and Smol, J.P. (eds). Kluwer Academic Publishers, Dordrecht.

Badeau, V., Becker, M., Bert, D., Dupouey, J.L., Lebourgeois, F., Picard, J.-F. (1996). Long-term growth trends of trees: Ten years of dendrochronological studies in France. In *Growth Trends in European Forests*, Spiecker, H., Mielikäinen, K., Köhl, M., Skovsgaard, J.P. (eds). Springer, Berlin, Heidelberg.

Beck, C. (2008). *Les eaux et forêts en Bourgogne ducale (vers 1350 – vers 1480) : société et biodiversité*. Éditions L'Harmattan, Paris.

Bergès, L. and Dupouey, J.-L. (2017). Écologie historique et ancienneté de l'état boisé : concepts, avancées et perspectives de la recherche. *Rev For Fr*, 69, 297–318.

Billamboz, A. (2014). Regional patterns of settlement and woodland developments: Dendroarchaeology in the Neolithic pile-dwellings on Lake Constance (Germany). *Holocene*, 24, 1278–1287.

Büntgen, U., Tegel, W., Nicolussi, K., McCormick, M., Frank, D., Trouet, V., Kaplan, J.O., Herzig, F., Heussner, K.-U., Wanner, H. et al. (2011). 2500 years of European climate variability and human susceptibility. *Sci.*, 331, 578–582.

Colombaroli, D. and Tinner, W. (2013). Determining the long-term changes in biodiversity and provisioning services along a transect from Central Europe to the Mediterranean. *Holocene*, 23, 1625–1634.

Conedera, M., Colombaroli, D., Tinner, W., Krebs, P., Whitlock, C. (2017). Insights about past forest dynamics as a tool for present and future forest management in Switzerland. *For Ecol Manage*, 388, 100–112.

Corvol, A. (ed.) (1999). *Les sources de l'histoire de l'environnement, le XIX*[e] *siècle*. Éditions L'Harmattan, Paris.

Costa, L., Laüt, L., Petit, C. (2020). Archéologie, forêt et LiDAR : une recherche qui a du relief. *Archéologies numériques*, 4(1), 1–7.

Devereux, B.J., Amable, G.S., Crow, P., Cliff, A.D. (2005). The potential of airborne lidar for detection of archaeological features under woodland canopies. *Antiquity*, 79(305), 648–660.

Dufraisse, A., Coubray, S., Girardclos, O., Nocus, N., Lemoine, M., Dupouey, J.-L., Marguerie, D. (2018). Anthraco-typology as a key approach to past firewood exploitation and woodland management reconstructions. Dendrological reference dataset modelling with dendro-anthracological tools. *Quat Int*, 463, 232–249.

Dupin, A., Girardclos, O., Fruchart, C., Laplaige, C. (2017). Anthracology of charcoal kilns in the forest of Chailluz (France) as a tool to understand Franche-Comte forestry from the mid-15th to the early 20th century AD. *Quat Int*, 458, 200–213.

Feiss, T., Horen, H., Brasseur, B., Buridant, J. (2017). Historical ecology of lowland forests: Does pedoanthracology support historical and archaeological data? *Quat Int*, 457, 99–112.

Fruchart, C. (2014). Analyse spatiale et temporelle des paysages de la forêt de Chailluz (Besançon, Doubs) de l'antiquité à nos jours. PhD Thesis, University of Franche-Comte, Besançon.

Georges-Leroy, M., Bock, J., Dambrine, É., Dupouey, J.-L. (2012). Les vestiges gallo-romains conservés dans le massif forestier de Haye (Meurthe-et-Moselle). Leur apport à l'étude de l'espace agraire. In *Des hommes aux champs. Pour une archéologie des espaces ruraux du Néolithique au Moyen Âge*, Carpentier, V. and Marcigny, C. (eds). Presses universitaires de Rennes, Rennes.

Girardclos, O. and Perrault, C. (2019). Les forêts de chêne du centre-est de la France. In *La forêt au Moyen Âge*, Richard, H. and Bepoix, S. (eds). Editions Les Belles Lettres, Paris.

Haneca, K., Wazny, T., Van Acker, J., Beeckman, H. (2005). Provenancing Baltic timber from art historical objects: Success and limitations. *J Archaeolog Sci*, 32, 261–271.

Iglesias, V., Vannière, B., Jouffroy-Bapicot, I. (2019). Emergence and evolution of anthropogenic landscapes in the western Mediterranean and adjacent atlantic regions. *Fire*, 2(4), 53.

Jouffroy-Bapicot, I. (2014). New palaeoecological data provided by the Premery forest closed depressions (Burgundy-France). *Quater*, 25(3), 253–269.

Ludemann, T., Michiels, H.-G., Nalken, W. (2004). Spatial patterns of past wood exploitation, natural wood supply and growth conditions: Indications of natural tree species distribution by anthracological studies of charcoal-burning remains. *Eur J For Res*, 123, 283–292.

Marquer, L., Gaillard, M.-J., Sugita, S. (2017). Quantifying the effects of land use and climate on Holocene vegetation in Europe. *Quat Sci Rev*, 171, 20–37.

Opitz, R.S. and Cowley, D.C. (eds) (2013). *Interpreting Archaeological Topography. Airborne Laser Scanning, 3D Data and Ground Observation*. Oxbow Books, Oxford.

Petrequin, P., Arbogast, R.M., Bourquin-Mignot, C., Lavier, C., Viellet, A. (1998). Demographic growth, environmental changes and technical adaptations: Responses of an agricultural community from the 32nd to the 30th centuries BC (Chalain, Clairvaux, archaeology). *W Archaeolog*, 30, 181–192.

Plue, J., Verheyen, K., Van Calster, H., Marage, D., Thompson, K., Kalamees, R., Jankowska-Blaszczuk, M., Bossuyt, B., Hermy, M. (2010). Seed banks of temperate deciduous forests during secondary succession. *J Veg Sci*, 21(5), 965–978.

Poublanc, S. (2015). Compter les arbres : une histoire des forêts méridionales à l'époque moderne. PhD Thesis, University of Toulouse-Jean Jaurès, Toulouse.

Renaux, B. and Villemey, A. (2016). Identifier et caractériser les forêts anciennes du Massif central. État des connaissances – boîte-à-outils – perspectives. Report, Conservatoire botanique national Massif-Central, Chavaniac-Lafayette.

Roberts, N., Fyfe, R.M., Woodbridge, J., Gaillard, M.-J., Davis, B.A.S., Kaplan, J.O., Marquer, L., Mazier, F., Nielsen, A.B., Sugita, S. et al. (2018). Europe's lost forests: A pollen-based synthesis for the last 11,000 years. *Sci Rep*, 8, 716.

Rochel, X. (2017). Une biogéographie historique. Forêts et industries dans le comté de Bitche au XVIIIe siècle. *Histoire et Mesure*, 32(2), 9–38.

Salomón, R., Rodríguez-Calcerrada, J., Zafra, E., Morales-Molino, C., Rodríguez-García, A., González-Doncel, I., Oleksyn, J., Zytkowiak, R., López, R., Miranda, J.C. et al. (2016). Unearthing the roots of degradation of *Quercus pyrenaica* coppices: A root-to-shoot imbalance caused by historical management? *For Ecol Manage*, 363, 200–211.

Saulnier, M., Cunill Artigas, R., Foumou, L.F. (2020). A study of late Holocene local vegetation dynamics and responses to land use changes in an ancient charcoal making woodland in the central Pyrenees (Ariège, France), using pedoanthracology. *Veget Hist Archaeobot*, 29, 241–258.

Seppä, H. (2007). Pollen analysis, principles. In *Encyclopedia of Quaternary Science*, Elias, S.A. (ed.). Elsevier Science, London.

Van Steenis, C. (1956). Basic principles of rain forest sociology. *Study of Tropical Vegetation. Proceedings of the Kandy Symposium, Ceylon*. UNESCO, Paris.

Watt, A.S. (1947). Pattern and process in the plant community. *J Ecol*, 35, 1–22.

Whittaker, R. (1953). A consideration of climax theory: The climax as a population and pattern. *Ecol Monogr*, 23(1), 41–71.

Zohary, D. and Hopf, M. (2000). *Domestication of Plants in the Old World: The Origin and Spread of Cultivated Plants in West Asia, Europe, and the Nile Valley*, 3rd edition. Oxford University Press, Oxford.

12

The Gate to the Forest is in its History

Keith J. KIRBY
University of Oxford, UK

12.1. Introduction

Trees, woodlands and forests can only be understood properly through their history. Individual trees may be several hundred years old; the growth rates shown in their annual rings reflect past changes in climate and levels of competition from nearby trees. The forms of trees often point to past management practices such as coppicing; earthworks such as boundary banks further emphasize previous human activity. The distribution of trees and woods in the landscape tells of land clearances, farm abandonments and forest creation.

Different approaches may be used to explore woodland history (Table 12.1; Watkins 2015). Changes in tree cover over thousands of years can be charted from pollen grains found at different depths in a soil or peat profile, patterns of fire from charcoal remains, and shifts in landscape openness from the species composition of subfossil beetles and snails. Archaeologists interpret the nature of woodland around settlements by analyzing the wood used in buildings, boats or for firewood. Woods were valuable properties and wood as a material was widely traded: historic documents may refer to trees as boundary markers; court reports tell how the ownerships were challenged and disputes over damage to the trees settled. Throughout the last 1,000 years, there have been exhortations and instructions as to how trees and woods should be managed from royal decrees to estate forestry records, albeit (as now) these were not always followed in practice. Early maps

Historical Ecology,
coordinated by Guillaume DECOCQ.

distinguished wooded from non-wooded lands because of the value of woodland, or the threat its presence might pose in harboring outlaws and enemy troops. This chapter explores how such knowledge from the past has also been used to improve woodland conservation.

Years BP	Source and information contained
10,000-2000	Tree composition from peat core (Hone *et al.* 2001)
c.1700	Roman pottery suggests farming in Wytham area
1070	Anglo-Saxon charter refers to woodland in Wytham area
900	Monastic records for woodland in Wytham area with hunting and firewood rights
c.500	Boundary banks (still recognisable) established around named coppices
	The oldest tree surviving in the area, the Broad Oak, starts to grow
480	Accounts of disposal of monastic land refers to named coppices in the Wytham area
400	Vice-chancellor's account books refer to Wytham Woods, as source of oaks bought by the University
260	First detailed map showing woodland boundaries (further maps through to present day)
	Accounts of wood and timber sales in local trade magazine
200	Estate papers describe major plantings and reorganisation of woods
	Ring counts show oaks surviving from this period
100	Aerial photograph shows a corner of the Woods with World War I training trenches
80	First student management plan written as part of forestry degree course
70	First complete aerial photographic cover for the Woods
	Charles Elton starts to record visits to Wytham in his diaries (Elton 1942-1965)
	First formal management plan produced by the University Forestry Department
65	Grayson and Jones produce their history of the Woods.

Table 12.1. *Types of evidence used to explore the history of Wytham Woods, near Oxford in southern England (Grayson and Jones 1955; Savill et al. 2010)*

12.2. The ancient woodland idea

A landowner in eastern England speculated that the presence of the bluebell *Hyacinthoides non-scripta* might indicate remnants of the "aboriginal" woodland cover (Beevor 1925). E.W. Jones, an ecologist in the University of Oxford, commented that the herb paris *Paris quadrifolia* was in those parts of Wytham Woods marked on a 1761 map, but not in adjacent plantations created on farmland in the early 19th century. In the 1970s, extensive historical and field work in Lincolnshire and East Anglia confirmed that some vascular plants were much more likely to be abundant in woodland patches that had existed for several hundred years or more (Peterken 1974; Rackham 2003). A similar association exists across much of Europe (Honnay et al. 1999; Hermy 2015).

Older woods tend to contain a wide range of interesting vascular plants and are also more likely to have other associated biological and cultural values (Peterken 1983). In Britain, the term "ancient woods" was adopted for sites that appeared to have a continuity of woodland cover for the last few hundred years, if not longer. A project was set up to list these across Britain using historical maps, documents and aerial photographs (Spencer and Kirby 1992; Goldberg et al. 2007, 2011; Figure 12.1). This produced a provisional inventory of ancient woods that has proved useful in prioritizing action both with respect to local site management and at a national level. Subsequent changes in forestry and urban policies mean that fewer ancient woods are now converted to conifer plantations or cleared for housing.

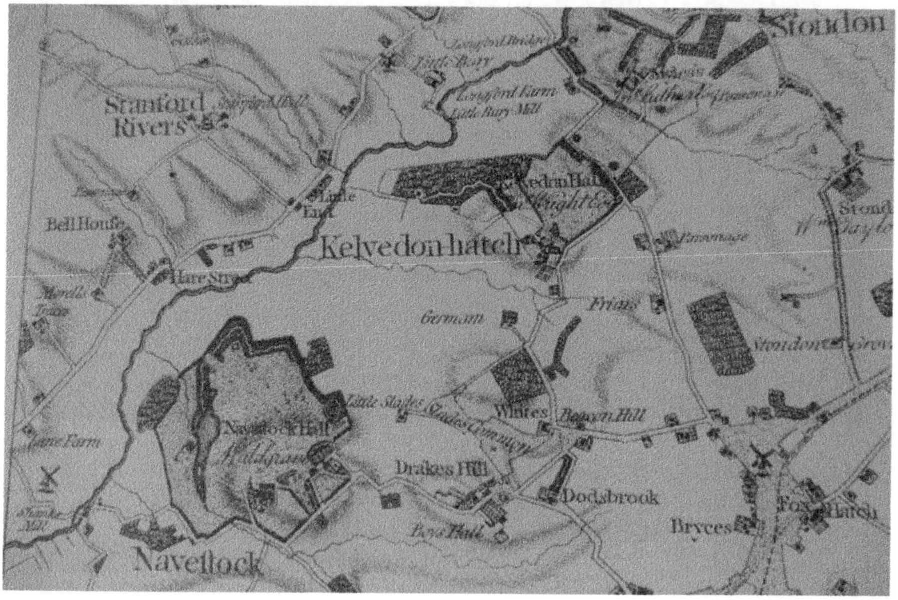

Figure 12.1. *Woodland shown on a map (c. 1:10,000 scale) from southern England (dated AD 1777); most of these patches are still present*

12.3. Legacies of woodland management

Rackham (1990, first published 1976) alerted the conservation sector to the importance of woodland management for maintaining features and species of interest. His work and studies inspired by him show how estate records and field work can be combined to reconstruct past woodland composition and structures (Figure 12.2). Oral history is also important in capturing ordinary, everyday practices. In Britain, just after World War II, many of the old crafts were dying out, but it was still possible to record the experience of the workers concerned (Edlin 1949). Grigson (1975) working slightly later (c. 1958) compiled lists of local plant names and their folk uses, which would not be possible today. Similarly, Bürgi and Gimmi (2007) used oral history sources to document the often-overlooked practice of litter collection. Recording the detail of past management is necessary if their effects are to be understood.

Figure 12.2. *Boundary bank and old coppice stools of* Carpinus betulus *for one of the woods in Figure 12.1. For a color version of this figure, see www.iste.co.uk/decocq/ecology.zip*

Despite increased protection for ancient woodland, many woodland species have continued to decline even in areas where the extent of ancient woodland remains high. Changes in historical woodland management, in particular the decline of coppicing, often underlie these losses of species that depend on open conditions or young growth stands (Buckley and Mills 2015a). In Britain, species affected include the butterfly *Boloria selene*, the bird *Luscinia megarhynchos* and the woodland plant *Carex depauperata*. Open conditions are less widespread under modern high forest management, because coppice rotations are much shorter than those for high forest (Buckley and Mills 2015b). Open woodland would also have allowed more frequent regeneration of oak, which is often only now present as large trees (Chudomelová et al. 2017). Less obvious changes in the ground flora may be happening as a result of the decline in collection of litter with the associated removal of nutrients from the woodland system (Dzwonko and Gawroński 2002).

12.4. Seeing the trees, not the woods

Foresters and woodland ecologists focus on woods and forests, but landscapes with scattered open-grown trees are common and were often more widespread in the past. Wood-pastures, where the ground between the trees was kept open by grazing

(Plieninger et al. 2015), are important for their vegetation mosaics and for large old trees, important for their saproxylic invertebrates, fungal populations and epiphytes (Hartel et al. 2015; Sebek et al. 2016). Their history can be more difficult to trace than that of closed woodland because open woodland might be mapped as closed woodland (if the trees were dense enough) or as open treeless ground in other circumstances. However, such sites were often associated with historic royal hunting areas and with parkland around castles or other large houses (Jonsell 2012; Farjon 2017). These are more likely to be well-represented in written records and appear in drawings and pictures, perhaps as background to the house or castle.

Individual trees may be recorded as geographic markers, as in Anglo-Saxon charters (Hooke 2010), Scandinavian path markers (Ericsson et al. 2003) or the witness trees used in land surveys by European settlers in North America (Whitney 1994). Ancient trees may acquire their own stories and traditions that can generate support for conservation interest and action in the real landscape history. For example, the Major Oak in Sherwood Forest attracts visitors because of its supposed association with Robin Hood. Other trees come to act as local or national symbols, such the Gernika oak that stood in front of the parliament of the Biscay Province in the Basque country (Leroy et al. 2020).

In parts of Europe, trees and shrubs themselves form the boundary of fields, from remnants left after woodland clearance to new lines planted or self-sown along fences. For the last 200 years, these boundaries have generally been shown on large-scale maps. Their creation or clearance in response to changing farming practice has a strong influence on the ecology of the intervening farmland. The hedge forms a habitat in its own right and may act as a corridor for species movement through the landscape (Watt and Buckley 1994; Davies and Pullin 2007).

12.5. Exploring the distant past

Human activity has been the dominant force shaping the distribution and character of trees and woods across much of Europe for the last few thousand years, but what were the more natural patterns that developed prior to the expansion of Neolithic farming? Can such landscapes be the baseline model for conservation (Hodder et al. 2009) or act as a guide for what might develop under future "rewilding" (Pettorelli et al. 2019)? Much of the potential natural vegetation of northwest Europe has been described as different types of largely closed forest (Bohn et al. 2000). However, an alternative view is that the landscape was quite open and more like a wood-pasture (Rackham 1998; Vera, 2000): this openness was maintained by populations of large, wild herbivores, particularly the wild ox *Bos primigenius* and the bison *Bison bonasus* (Table 12.2).

Predominantly wooded landscape, cf. Peterken (1996)	Half-open wood-pasture type landscape cf. Vera (2000)
Temporary and permanent open space present in a matrix of closed canopy stands. Openness probably less than 20% of the total area.	Half-open landscape with much scrub, scattered trees and stands of trees, 'groves' which might be up to tens or a few hundreds of hectares across.
Closed canopy areas tended to maintain themselves by regeneration in gaps within the stand.	Very limited regeneration within stands, which eventually broke down to open ground; regeneration to scrub and tree cover concentrated in the currently open stands.
Large herbivores present, but not the main shapers of the landscape structure. Regeneration patterns generally driven by other factors (wind, disease, fire, flood, and early human activity, etc).	Large herbivores the main drivers of landscape structure and composition. Other drivers much less significant.
Shade tolerant trees such as beech and small-leaved lime became abundant; oak remained present, but a more minor component.	Oak remained abundant; succession to more shade-tolerant species was limited by high grazing pressure.
Development of open managed woodland represents a shift from the 'natural state'.	Abundance of oaks maintained by grazing; shift to managed woodland and wood-pasture maintained an analogue of the natural landscape.
Rewilding might be expected to lead to an increase in tree cover, tending towards largely-closed landscapes	Rewilding should include sufficient large herbivores to produce shifting mosaics of open, scrub and closed grove vegetation.

Table 12.2. *Contrasting visions for European natural landscapes in the mid-Holocene*

Palynological studies favor a more closed landscape with little evidence for the extensive grassland or heathland appearing in the pollen record (Mitchell 2005). Where there is an increase in indicators of open landscapes, this is generally associated with evidence of climatic change or human intervention (Nielsen et al. 2012; Fyfe et al. 2013).

Analysis of subfossil beetle remains provides evidence for a more open landscape: many of the species detected depend currently on old open-grown oaks in relatively sunny situations (Sebek et al. 2016). However, it is not clear how important grazing animals were in creating and maintaining open areas (Whitehouse and Smith 2010; Sandom et al. 2014). Robinson (2014) suggests that something like Vera's proposed shifting mosaic may have developed following the clearance and land abandonment that likely characterized early Neolithic farming. Allen (2017)'s analysis of snail faunas also suggests a link between open landscapes and high human use, rather than necessarily the activity of wild herbivores.

An unsolved problem is finding a reliable proxy for the density of large herbivores. The spores of fungi that live on dung can be extracted from peat and soil cores, but, as yet, there is not a reliable way of linking spore abundance to animal numbers (Davies 2019). Nor is it clear that a half-open landscape would necessarily result from the impacts of large herbivores alone (Kirby 2004).

12.6. Trees and woods from the past to the future

The binary division between ancient and recent woodland has proved useful for assigning broad conservation priorities, but historical ecology shows that there is more of a continuum in woodland characteristics. Dupouey et al. (2002) highlight that legacies of Roman-era clearance may still be detectable in present-day woodland soils in some French forests. Barnes and Williamson (2015) report ancient woodland indicators common in some woods that from documentary evidence are clearly recent. Saproxylic insects, lichens and bryophytes found in or on old trees show that some aspects of "woodland ecosystems" do not depend on being in closed woodland at all (Figure 12.3).

Figure 12.3. *Many saproxylic invertebrates and epiphytic lichens depend on old trees, but not on woodland. For a color version of this figure, see www.iste.co.uk/decocq/ecology.zip*

Changing woodland management, or stopping it completely, as in strict forest reserves and under rewilding has significant impacts on the legacy of species and assemblages that we inherited (Peterken and Mountford 2017). Moreover, future climatic and atmospheric conditions will be different from those in which ancient trees and woods developed. We cannot study the effects of future changes without knowing how trees and woods arrived at the present (Burgi et al. 2020). We should use the insights from historical ecology to help conserve the meaning in woods from the past and to develop new meanings for future woods and treescapes.

12.7. References

Allen, M.J. (2017). The southern English chalklands: Molluscan evidence for the nature of the post-glacial woodland cover. In *Molluscs in Archaeology: Methods, Approaches and Applications*, Allen, M.J. (ed.). Oxbow Books, Oxford, UK.

Barnes, G. and Williamson, T. (2015). *Rethinking Ancient Woodland: The Archaeology and History of Woods in Norfolk*. University of Hertfordshire Press, Hatfield, UK.

Beevor, H. (1925). Norfolk woodlands from the evidence of contemporary chronicles. *Quart. J. Forest.*, 19, 87–110.

Bohn, U., Gollub, G., Hettwer, C. (2000). *Map of the Natural Vegetation of Europe*. Federal Agency for Nature Conservation, Bonn, Germany.

Buckley, G.P. and Mills, J. (2015a). Coppice silviculture: From the mesolithic to 21st century. In *Europe's Changing Woods and Forests: From Wildwood to Managed Landscapes*, Kirby, K.J. and Watkins, C. (eds). CABI, Wallingford, UK.

Buckley, G.P. and Mills, J. (2015b). The flora and fauna of coppice woods: Winners and losers of active management or neglect. In *Europe's Changing Woods and Forests: From Wildwood to Managed Landscapes*, Kirby, K.J. and Watkins, C. (eds). CABI, Wallingford, UK.

Bürgi, M. and Gimmi, U. (2007). Three objectives of historical ecology: The case of litter collecting in Central European forests. *Land. Ecol.*, 22, 77–87.

Bürgi, M., Cevasco, R., Demeter, L., Fescenko, A., Gabellier, N., Marull, J., Ostlund, L., Santruchkova, M., Wohlgemuth, T. (2020). Where do we come from? Cultural heritage in forests and forest management. In *How to Balance Forestry and Biodiversity Conservation. A View Across Europe*, Krumm, F., Schuck, A., Rigling, A. (eds). EFI, WSL, Bern, Switzerland.

Chudomelová, M., Hédl, R., Zouhar, V., Szabó, P. (2017). Open oakwoods facing modern threats: Will they survive the next fifty years? *Biol. Conserv.*, 210, 163–173.

Davies, A.L. (2019). Dung fungi as an indicator of large herbivore dynamics in peatlands. *Rev. Palaeobot. Palynol.*, 271, 104–108.

Davies, Z.G. and Pullin, A.S. (2007). Are hedgerows effective corridors between fragments of woodland habitat? An evidence-based approach. *Landsc. Ecol.*, 22, 333–351.

Dupouey, J.L., Dambrine, E., Laffite, J.D., Moares, C. (2002). Irreversible impact of past land use on forest soils and biodiversity. *Ecology*, 83, 2978–2984.

Dzwonko, Z. and Gawroński, S. (2002). Effect of litter removal on species richness and acidification of a mixed oak-pine woodland. *Biol. Conserv.*, 106, 389–398.

Edlin, H.L. (1949). *Woodland Crafts in Britain*. Batsford, London, UK.

Elton, C.S. (1942–1965). Elton Archive: Notes on the Wytham Area 1942–1965 [Online]. Available at: http://ora.ox.ac.uk/objects/uuid:89c5e479-6003-45ba-bd78-8a8a12858bf1.

Ericsson, T.S., Ostlund, L., Andersson, R. (2003). Destroying a path to the past – The loss of culturally scarred trees and change in forest structure along Allmunvagen, in mid-west boreal Sweden. *Silva Fennica*, 37, 283–298.

Farjon, A. (2017). *Ancient Oaks in the English Landscape*. Kew, London.

Fyfe, R.M., Twiddle, C., Sugita, S., Gaillard, M.-J., Barratt, P., Caseldine, C.J., Dodson, J., Edwards, K.J., Farrell, M., Froyd, C. et al. (2013). The Holocene vegetation cover of Britain and Ireland: Overcoming problems of scale and discerning patterns of openness. *Quat. Sci. Rev.*, 73, 132–148.

Goldberg, E.A., Kirby, K.J., Hall, J.E., Latham, J. (2007). The ancient woodland concept as a practical conservation tool in Great Britain. *J. Nat. Conserv.*, 15, 109–119.

Goldberg, E.A., Peterken, G.F., Kirby, K.J. (2011). Origin and evolution of the ancient woodland inventory. *B. Wildl.*, 23, 90–96.

Grayson, A.J. and Jones, E.W. (1955). *Notes on the History of the Wytham Estate with Special Reference to the Woodlands*, Imperial Forestry Institute, Oxford, UK.

Grigson, G. (1975). *The Englishman's Flora*. Paladin, St Albans, UK.

Hartel, T., Plieninger, T., Varga, A. (2015). Wood-pastures in Europe. In *Europe's Changing Woods and Forests: From Wildwood to Managed Landscapes*, Kirby. K.J. and Watkins, C. (eds). CABI, Wallingford, UK.

Hermy, M. (2015). Evolution and changes in the understorey of deciduous forests: Lagging behind drivers of change. In *Europe's Changing Woods and Forests: From Wildwood to Managed Landscapes*, Kirby. K.J. and Watkins, C. (eds). CABI, Wallingford, UK.

Hodder, K.H., Buckland, P.C., Kirby, K.J., Bullock, J.M. (2009). Can the pre-Neolithic provide suitable models for re-wilding the landscape in Britain? *British Wildlife (supplement)*, 20, 4–15.

Hone, R., Anderson, D.E., Parker, A.G., Morecroft, M.D. (2001). Holocene vegetation change at Wytham Woods, Oxfordshire. *Quat. Newsletter*, 94, 1–15.

Honnay, O., Hermy, M., Coppin, P. (1999). Effects of area, age and diversity of forest patches in Belgium on plant species richness, and implications for conservation and reforestation. *Biol. Conserv.*, 87, 73–84.

Hooke, D. (2010). *Trees in Anglo-Saxon England: Literature, Lore and Landscape*. The Boydell Press, Woodbridge, UK.

Jonsell, M. (2012). Old park trees as habitat for saproxylic beetle species. *Biodiv. Conserv.*, 21, 619–642.

Kirby, K.J. (2004). A model of a natural wooded landscape in Britain as influenced by large herbivore activity. *Forestry*, 77, 405–420.

Leroy, T., Plomion, C., Kremer, A. (2020). Oak symbolism in the light of genomics. *New Phytol.*, 226, 1012–1017.

Mitchell, F.J.G. (2005). How open were European primeval forests? Hypothesis testing using palaeoecological data. *J. Ecol.*, 93, 168–177.

Nielsen, A.B., Giesecke, T., Theuerkauf, M., Feeser, I., Behre, K.E., Beug, H.J., Chen, S.H., Christiansen, J., Dörfler, W., Endtmann, E. et al. (2012). Quantitative reconstructions of changes in regional openness in north-central Europe reveal new insights into old questions. *Quat. Sci. Rev.*, 47, 131–149.

Peterken, G.F. (1974). A method for assessing woodland flora for conservation using indicator species. *Biol. Conserv.*, 6, 239–245.

Peterken, G.F. (1983). Woodland conservation in Britain. In *Conservation in Perspective*, Warren, A. and Goldsmith, F.B. (eds). John Wiley & Sons, Chichester, UK.

Peterken, G.F. (1996). *Natural Woodland: Ecology and Conservation in Northern Temperate Regions*. Cambridge University Press, Cambridge, UK.

Peterken, G.F. and Mountford, E.P. (2017). *Woodland Development: A Long-term Study of Lady Park Wood*. CABI, Wallingford, UK.

Pettorelli, N., Durant, S.M., Du Toit, J.T. (2019). *Rewilding*. Cambridge University Press, Cambridge, UK.

Plieninger, T., Hartel, T., Martín-López, B., Beaufoy, G., Bergmeier, E., Kirby, K., Montero, M.J., Moreno, G., Oteros-Rozas, E., Van Uytvanck, J. (2015). Wood-pastures of Europe: Geographic coverage, social–ecological values, conservation management, and policy implications. *Biol. Conserv.*, 190, 70–79.

Rackham, O. (1990). *Trees and Woodland in the British Landscape (Revised Edition)*. Dent, London, UK.

Rackham, O. (1998). Savanna in Europe. In *The Ecological History of European Forests*, Kirby, K.J. and Watkins, C. (eds). CABI, Wallingford, UK.

Rackham, O. (2003). *Ancient Woodland: Its History, Vegetation and Uses in England (Revised Edition)*. Castlepoint Press, Dalbeattie, UK.

Robinson, M. (2014). The ecodynamics of clearance in the British Neolithic. *Env. Archaeol.*, 19, 291–297.

Sandom, C.J., Ejrnæs, R., Hansen, M.D.D., Svenning, J.C. (2014). High herbivore density associated with vegetation diversity in interglacial ecosystems. *Proc. Nat. Acad. Sci. USA*, 111, 4162–4167.

Savill, P.S., Perrins, C., Kirby, K.J., Fisher, N. (2010). *Wytham Woods, Oxford's Ecological Laboratory*. Oxford University Press, Oxford, UK.

Sebek, P., Vodka, S., Bogusch, P., Pech, P., Tropek, R., Weiss, M., Zimova, K., Cizek, L. (2016). Open-grown trees as key habitats for arthropods in temperate woodlands: The diversity, composition, and conservation value of associated communities. *For. Ecol. Manage.*, 380, 172–181.

Spencer, J.W. and Kirby, K.J. (1992). An inventory of ancient woodland for England and Wales. *Biol. Conserv.*, 62, 77–93.

Vera, F.W.M. (2000). *Grazing Ecology and Forest History*. CABI, Wallingford, UK.

Watkins, C. (2015). Methods and approaches in the study of woodland history. In *Europe's Changing Woods and Forests: From Wildwood to Managed Landscapes*, Kirby, K.J. and Watkins, C. (eds). CABI, Wallingford, UK.

Watt, T.A. and Buckley, G.P. (1994). *Hedgerow Management and Nature Conservation*. Wye College Press, Ashford, UK.

Whitehouse, N.J. and Smith, D. (2010). How fragmented was the British Holocene wildwood? Perspectives on the "Vera" grazing debate from the fossil beetle record. *Quat. Sci. Rev.*, 29, 539–553.

Whitney, G.G. (1994). *From Coastal Wilderness to Fruited Plain: A History of Environmental Change in Temperate North America, 1500 to the Present*, Cambridge University Press, Cambridge, UK.

13

Plant Assemblages and Ecosystem Functioning, a Legacy of Long-term Interactions with Large Herbivores

Christophe Baltzinger and Anders Mårell
INRAE, UR EFNO, Nogent-sur-Vernisson, France

13.1. Introduction

Large herbivores are primary consumers, intermediates in the trophic web between primary producers and predators. They lie at the center of a rich network of direct and indirect interactions, both biotic (as prey and with organisms depending on similar feeding resources) and abiotic (with different compartments shaping their home ranges). They intervene in several ecological processes from home range to landscape to regional scale and thereby affect assembly rules in plant communities and ecosystem functioning. In this chapter, we will first highlight how large herbivores shape local plant assemblages, as ecosystem dominant interactors. We will then unveil how ecologists overcame methodological challenges and made it possible to monitor the effects of large herbivores at different spatial scales over the long-term. We also summarize the hypotheses related to how they shaped the ecosystems during the Late Pleistocene and Holocene, and reveal their role in more

Historical Ecology,
coordinated by Guillaume Decocq.

current temporal vegetation dynamics, including legacy effects, at the population and community levels. In conclusion, we mention passive and active rewilding dynamics, and the role large herbivores play in species and ecosystem conservation and in degraded ecosystem restoration in a context of global change.

13.2. Large herbivores are ecosystem dominant interactors

Large herbivores include more than 250 wild, introduced and domestic (i.e. livestock) species, some of which are ruminants (e.g. bovidae, cervidae) and some of which are not (e.g. equidae). Their respective feeding strategies also vary along a gradient from grazers (i.e. grass and roughage eaters) to browsers (i.e. concentrate selectors) depending on their digestive morphology and physiology. Intermediate mixed feeders, like the red deer (*Cervus elaphus*), lie in between, feeding on grass and roughage as well as on richer items (e.g. leaves, buds). The feeding strategy, feeding selectivity, body mass and abundance require attention when assessing the effects that large herbivores have on plant communities.

13.2.1. *Large herbivores as ecosystem engineers*

Besides their trophic interactions, large herbivores are ecosystem engineers (i.e. "[...] organisms that directly or indirectly modulate the availability of resources to other species, by causing physical state changes in biotic or abiotic materials. In so doing they modify, maintain and create habitats [...]", Jones et al. 1994, Figure 13.1). Wilby et al. (2001) included trophic interactions in the concept of ecosystem engineering, because it is sometimes difficult to disentangle trophic interactions from engineering effects (e.g. wild boar rooting in the soil for bulbs and rhizomes). Herbivores are involved in intrinsic trophic interactions that can affect the physical state of the woody understory vegetation (i.e. shrubs and young trees) and its attributes as habitat for other organisms to hide from predators, provide thermal comfort, nest or feed. They directly consume selected plants and indirectly affect neglected ones. They also compete with other herbivores relying on similar resources, for example, caterpillars, on which birds depend to feed their chicks (i.e. indirect trophic cascading effects). Large herbivores can also predate non-plant life forms (e.g. earthworms, insects, eggs; Cocquelet et al. 2019). They also affect soil characteristics, through physical engineering (deep or surface disturbance) that affects fluxes in the soil, chemical (e.g. nutrients, water) and physical (e.g. mud, diaspore[1]) transport.

1. A spore, seed or other structure that allows plant dispersal.

Figure 13.1. *The wisent (Bison bison), a keystone species, involved in diverse ecological ecosystem processes (e.g. fur-epizoochory, trampling, herbivory, nutrient cycling), has been reintroduced back to the Southern Carpathians rewilding area. Artwork by Jeroen Helmer/ARK Nature www.ark.eu/wisent. For a color version of this figure, see www.iste.co.uk/decocq/ecology.zip*

13.2.2. *Large herbivores and plant assemblages*

Local plant assemblages are shaped from the plant species belonging to the regional species pool. These plants must undergo three main filters to integrate the extant plant community (Lortie et al. 2004). Sequentially, we first have dispersal (i.e. ability to reach a new environment), then the environmental filter (i.e. local abiotic conditions plants have to tolerate) and finally, the biotic context (i.e. interactions with other plants or other types of organisms). Davis et al. (2000) detail how fluctuating resources in local plant communities can explain how newcomers find opportunities to enter a community. These opportunities depend on diaspore arrival, the characteristics of the new plant and the susceptibility of the local environment to being colonized, i.e. the environment's ability to provide transient resources (i.e. released locally and/or added from the outside) for plants to establish and grow.

Large herbivores are involved in each of these filters. They help direct long-distance dispersal for many plants via endozoochory and epizoochory (Baltzinger et al. 2019). Dispersed plants bear specific traits (e.g. diaspore appendage or high releasing height favor epizoochory); they are usually released in areas with similar habitat conditions. Endozoochory also influences the build-up of the soil seed bank (e.g. Jaroszewicz 2013 for wisent). More than just modifying the local environment, large herbivores drive resource heterogeneity (e.g. nutrients, light and water) and create microhabitats in relation to their habitat selection patterns. Nutrient deposition in feces/urine and decomposing carcasses create long-lasting biogeochemical hotspots (Murray et al. 2013). Small-scale disturbances (e.g. litter scraping, trampling, feeding-associated digging and uprooting) affect soil physical structure, chemical composition and mesofauna (Mohr et al. 2005). Selective herbivory and stem breaking (Olmsted et al. 2021) also modify understory vertical structure, affecting its ability to filter light and delaying forest succession. This cascades on organisms like arachnids or birds, which depend on specific vegetation structures for web-building or nesting. Selective herbivory also mediates plant–plant interactions via apparent competition. Floral herbivory (Mårell et al. 2009) may also affect plant–pollinator interactions. Endozoochory influences plant germination ability and timing (Milotić and Hoffmann 2016). Biotic interactions also include secondary dispersal by dung beetles, predation on diaspores concentrated in feces by rodents or decomposition by fungi developing on feces that modulate dispersal effectiveness.

Though each of these three main filters shapes local plant assemblages, escape from strong biotic context (i.e. intense chronic herbivory for 20 years) highlighted the environmental filter acting on forest plant assemblages (Chollet et al. 2021).

13.3. Long-term effects and methodological changes

13.3.1. *Paleoecological records*

Reconstructing plant–herbivore interactions in paleoecological times requires specific approaches (e.g. Bradshaw et al. 2003). One approach is to associate scattered remains found in similar-aged layers. Radiocarbon dating can provide accurate estimations for the age of the matrix, or direct dates for bony remains more than 30,000 years (1,000 years = KY, hereafter) old. Ancient vegetation remains contain plant macrofossils (i.e. seeds, fruits and leaves) where the plants initially grew, but pollen, another important indicator, may have been dispersed over long distances, and should be interpreted at larger spatial scales. Herbivore remains (bones and teeth) are found more rarely, and their occurrence does not indicate abundance. Often, herbivore remains are retrieved from the same sediment layers as vegetation remains. Tooth wear can also help determine what large herbivores consumed during ancient times. Abundance and identity of dung beetles, coprophilous fungal spores and pollen records can inform about the guild of large herbivores and past grazing regimes. Archaeological digs and cave deposits represent main sites of interest combining archaeozoology, anthracology and paleoanthropology (Defleur and Desclaux 2019).

13.3.2. *Modern data*

More recently, several methods have been applied to plant–herbivore interactions. They generally require the spatial comparison of two different levels of herbivore abundance. For example, presence and absence could be compared by experimentally excluding ungulates from a given area (e.g. Bernes et al. 2018) or by investigating natural refuges (e.g. rocky outcrops, Chollet et al. 2013). Areas hosting various herbivore densities can also be compared via data on different levels of hunting pressure (Boulanger et al. 2015), accessibility (Martin and Baltzinger 2002), eradication attempts, drastic population reductions (Tanentzap et al. 2009) or management options combining fencing and supplemental feeding (Baltzinger et al. 2016). Scientists have also simulated herbivory by clipping vegetation at different periods of the year (Laurent et al. 2017). In other cases, herbivore impact has been simulated by jointly clipping vegetation and releasing nitrogen. Persson et al. (2005) used this strategy to reflect four levels of moose (*Alces alces*) populations (0, 0.8, 2.4 or 4 ind.km^{-2}). However, their experimental setup could have been more complete by considering trampling. Tremblay et al. (2006) also manipulated herbivore density by introducing a known number of individuals into fenced compounds with predetermined surface areas.

The types of comparisons described above are static, and assessing temporal changes in plant–herbivore interactions requires longitudinal approaches, for example, permanent plots resurveyed at different time scales (from seasons to years to decades, for example, forestREplot[2]) for plant communities and herbivory pressure (e.g. browsing index, abundance assessment, Boulanger et al. 2015). However, the longer the time period considered, the greater the risk of observer effect and plot relocation issues. Abundance for herbivores outside exclosures is difficult to tackle and prone to annual changes (but see Boulanger et al. 2018). One interesting example is the synchronic studies in the Haida Gwaii archipelago, which took advantage of delayed colonization patterns (i.e. some islands have never been colonized, some were colonized 20 years ago and others more than 50 years ago; Stockton et al. 2005). This colonization history has been unveiled by dendrochronology, dating the last regeneration of palatable trees and the oldest fraying scars. The uncolonized islands serve as a reference state, making it possible to test for resilience. Chronosequences in combination with delayed exclusion experiments have also been used (e.g. Hidding et al. 2013) to test for legacy effects (i.e. "the ghost of herbivory past"). Finally, time series of aerial photographs can help study the effects of herbivores on forest succession.

Furthermore, different vegetation metrics (e.g. abundance in terms of cover percentage/number of individuals, height, morphology, phenology) have been usually monitored for various plant species in various treatments to make comparisons at the species, subgroup (e.g. woody vegetation, forbs, graminoids, bryophytes) and community levels (e.g. species richness, diversity indices) possible.

13.4. Plant–herbivore interactions over the long-term

13.4.1. *Quaternary communities of large herbivores and associated flora*

The Quaternary period includes the Pleistocene (2.588 million years to 11.7 KY before present, BP) followed by the Holocene lasting until the present. In Europe, this period presents animal remains of roughly 10–12 large herbivores from the late Pleistocene until the last interglacial period (128–114 KY BP), including mega herbivores, like giant deer, auroch, straight-tusked elephants (Bradshaw et al. 2003).

2. https://forestreplot.ugent.be/.

Habitat types are sometimes inferred from the supposed feeding strategy of these past herbivores. For instance, the penultimate interglacial period (240 KY BP) hosted a diverse fauna of grazers indicating savanna-like vegetation, whereas the latest interglacial period had fewer grazers suggesting more forested habitats. The early Holocene, just after the latest glaciation, hosted a depleted community of herbivores, which were smaller (Gill 2014). This transition also marks a shift from more grazing to more browsing herbivore communities, which are less prone to limit abundant grassland/savanna-like vegetation, with potential consequences on the occurrence and intensity of natural fires. Although bone remnants and animal imprints allow the identification of herbivore species, precise knowledge of their abundance remains challenging (but see section 13.3.1 for the use of herbivore-dependent organisms). This makes the understanding of past plant–herbivore interactions more difficult.

13.4.2. *The forest in the early Holocene*

There is a rich ongoing debate about what the forest in Northwestern Europe looked like after the last Ice Age in the early Holocene (Bradshaw et al. 2003). Vera (2000) proposed the "wood-pasture hypothesis", which suggests that large herbivores maintained a mosaic of open habitats, allowing light-demanding trees (oaks *Quercus* and hazel *Corylus*, present in the pollen records) and shrubs to persist within the forest. Under this hypothesis, large herbivores were the main agents of the open habitats and unpalatable or less consumed shrubs served as a physical protection for young, palatable trees during regeneration. However, Mitchell (2005) found high pollen records for oaks and hazel in Ireland in the absence of large herbivores, thus invalidating the idea that they played a key role in maintaining these tree species. Alternatively, the "high forest hypothesis" suggests that the early Holocene was predominantly composed of closed forests. Paleorecords for various taxa (e.g. tree pollen, fossils, insects and mollusks) support this hypothesis. However, we do not know how light-demanding species like oaks and hazel regenerated. Subsequently, Svenning (2002) suggested that there was probably a wide range of habitats and that open vegetation was more frequent on floodplains and infertile soils. Uplands would have been mostly closed forests, with some openings maintained (but not necessarily created) by large herbivores. It is further supposed that natural fire regimes, either favored or hampered by herbivores, shaped the diversity of the habitats in pre-agricultural landscapes. For instance, Waldram et al. (2008) showed that grazing by the white rhinoceros in rich mesic savannas controlled vegetation height and facilitated access for other herbivores, thus limiting the size and intensity of natural fires.

13.5. Modern vegetation trajectories driven by large herbivores

13.5.1. *Herbivory effects*

Bernes et al. (2018) recently performed a systematic review of ungulate herbivory on understory vegetation in temperate and boreal forests (i) at the community level (abundance and species richness), (ii) at the sub-group level (the woody understory – shrubs, seedlings and saplings – graminoids, forbs and bryophytes) and (iii) at the species level, whenever feasible. They also took into account the herbivore origin (domestic, introduced and native) and their feeding strategy (grazer, intermediate mixed feeder and browser).

Bernes et al.'s work points out the negative effects of ungulate herbivory on the abundance and species richness of woody understory vegetation and on bryophyte abundance, and the positive effects on forb and bryophyte species richness. Both abundance and height of most tree saplings tended to be limited by herbivory (e.g. *Quercus*, *Tsuga*), and this occurred regardless of the degree of palatability. A single forb (*Alliaria petiolata*), exotic and invasive in North America, benefitted from ungulate herbivory; whereas the abundance of three species was reduced (*Maianthemum canadense*, *Calluna vulgaris* and *Vaccinium vitis-idaea*) in their native range.

Ungulate origin also significantly affected vegetation patterns, both in abundance and composition. Domestic and introduced ungulates most depleted global understory abundance, though native ungulates had a noticeable impact on the woody understory. Only domestic herbivores (all grazers in this review) limited the abundance of graminoids. If we consider species richness, domestic ungulates favored global understory species richness, specifically forbs, and to a lesser extent graminoids, whereas native and introduced ungulates reduced woody understory species richness. The feeding strategy also had an effect; intermediate mixed feeders and grazers both reduced understory cover, whereas only grazers favored species richness.

13.5.2. *Temporal trajectories*

Bernes et al. (2018) also concluded that the current lack of replicated studies makes it difficult to assess the temporal effects of different types of herbivores over the long-term (i.e. six years was the median herbivory exposure length). The higher the herbivory intensity (i.e. herbivore abundance x exposure length), the lower the abundance for global understory vegetation, the woody understory and shrubs.

Sapling species richness also decreased with increased herbivory intensity, while forb species richness and graminoid abundance increased.

Long-term studies generally report trends in plant community changes (e.g. Ellenberg indicator values for light and nitrogen; dispersal modes; native status; Grime CSR strategies) or winner/loser species (i.e. increasing/decreasing in the frequency of occurrence) and their associated traits. Vild et al. (2017) took advantage of vegetation resampling inside and outside a game preserve, where herbivore populations had increased from 5 to 55 ind.km^{-2}. They found an increase in global species richness (see also Boulanger et al. 2018) linked to an increase in ruderal plants, mostly nitrophilous and light-demanding species, at the expense of woody and endangered species. Endozoochorous plants were generally favored, but not epizoo- or anemochorous species. However, dispersal syndromes are not reliable, as they underestimate zoochorous dispersal by ungulates (Green et al. 2022). Over a 50-year period, Wiegmann and Waller (2006) identified 21 winner and 21 loser plant species out of a total of 73 species. The winners, 16 natives and five exotics, were common species, mostly grasses and sedges, all with abiotic modes of dispersal and pollination, and all browsing-tolerant or browsing-resistant species. The 21 losers were rarer, deer-sensitive (palatable) species. Globally, the number of species per surface unit decreased over time, the winner species showing a more spatially consistent pattern and the loser species being more prone to random disappearances.

Legacy effects have been revealed in different studies, proving that plant communities maintain a memory of the previous herbivory events that have shaped their trajectories. Hidding et al. (2013) showed that only eight years of white-tailed deer (*Odocoileus virginianus*, hereafter WTD) herbivory after logging changed the plant community's early-successional trajectory. Furthermore, Pendergast et al. (2016) highlighted a lag time in plant community response following WTD exclusion: species composition inside exclusion fences started to diverge from the unfenced control plots only after five years. During a combined plant and herbivory resurvey of a 30-year monitoring, Boulanger et al. (2015) showed that the plant community changes were correlated with the gradient of deer herbivory pressure. However, the magnitude of these changes was lower than expected, again suggesting a delay in the response of the understory to relaxed deer herbivory pressure. Royo et al. (2010) tested the resilience of over-browsed understory vegetation after an experimental culling experiment. Few deer-palatable species responded quickly to the reduction in WTD populations, but species richness in the understory did not fully recover; browsing-resistant species continued to dominate, forming recalcitrant understory layers. In the absence of plant refuges providing a diaspore pool (Chollet et al. 2013), species that have been reduced to sparse or weak individuals will barely, or never, recover. Nuttle et al. (2014) also stressed the long-lasting legacy effects that continued for at least 20 years in the Northeastern US and concluded that deer population reduction alone was not sufficient for plant community recovery. Additional measures, such as constraining recalcitrant fern layers, must be taken. Tanentzap et al. (2009) reached similar conclusions in a context where introduced red deer populations had been drastically reduced (by 92%) over 40 years. Saatkamp et al. (2021) also evidenced very long-term sheep-pastoralism legacies in Southeastern France, leading to very high plant species richness.

13.6. Perspectives, rewilding and ecosystem restoration

Large herbivores have shaped past and present landscapes through a diversity of ecological processes, among which herbivory is not the least. Currently, a recovery process of wild ungulates (mostly browsers) is taking place in Europe and populations will probably reach unprecedented abundance levels. This is being accompanied by the passive return of their natural predators (e.g. lynx, wolf), thus restoring natural complexity to some extent in current landscapes. Herbivores, as prey, will probably see their behavior and population dynamics affected, and vegetation abundance and diversity will certainly change through cascading effects. A recent worldwide assessment of the ecological functions fulfilled by contemporary introduced herbivores shows that they partly restore vanished ecological functions previously provided by ancient herbivores (before the late Pleistocene extinctions, Lundgren et al. 2020). Current rewilding programs aim to bring back populations of grazing animals (e.g. wisent in Romania[3], Figure 13.1) capable of maintaining open habitats and safeguarding the associated faunal and floral diversity (Garrido et al. 2018). Large herbivores create heterogeneity, not only in their home ranges but also at the landscape scale, and may be dynamic actors in restoring degraded habitats (Lundberg and Moberg 2003) through the key interactive ecological processes bridging plants and animals (e.g. herbivory, diaspore and nutrient transport, physical engineering). Boulanger et al. (2018) demonstrated that the presence of large herbivores extended the survival of early successional plants under forest cover; Rico et al. (2014) showed that rotational shepherding favored the dispersal of non-specialized diaspores. Large herbivores might not only be of help in restoring degraded habitats, but they might also improve the conservation of species and ecosystems over the long-term. However, given the natural dynamics involving large herbivores, active rewilding programs in human-dominated landscapes must be coupled with effective monitoring of diverse taxa (e.g. native and exotic flora, entomofauna and avifauna). This will ensure expected ecological outputs and social

3. https://rewildingeurope.com/areas/southern-carpathians/.

acceptance requirements are met in order to fully support the current UN decade on ecosystem restoration 2021–2030[4].

13.7. References

Baltzinger, M., Mårell, A., Archaux, F., Pérot, T., Leterme, F., Deconchat, M. (2016). Overabundant ungulates in French Sologne? Increasing red deer and wild boar pressure may not threaten woodland birds in mature forest stands. *Basic and Applied Ecology*, 17(6), 552–563.

Baltzinger, C., Karimi, S., Shukla, U. (2019). Plants on the move: Hitch-hiking with ungulates distributes diaspores across landscapes. *Frontiers in Ecology and Evolution*, 7(38), 1–19. doi: 10.3389/fevo.2019.00038.

Bernes, C., Macura, B., Jonsson, B.G., Junninen, K., Müller, J., Sandström, J., Lõhmus, A., Macdonald, E. (2018). Manipulating ungulate herbivory in temperate and boreal forests: Effects on vegetation and invertebrates. A systematic review. *Environmental Evidence*, 7(13), 1–32.

Boulanger, V., Baltzinger, C., Saïd, S., Ballon, P., Picard, J.-F., Dupouey, J.-L. (2015). Decreasing deer browsing pressure influenced understory vegetation dynamics over 30 years. *Annals of Forest Science*, 72(3), 367–378.

Boulanger, V., Dupouey, J.-L., Archaux, F., Badeau, V., Baltzinger, C., Chevalier, R., Corcket, E., Dumas, Y., Forgeard, F., Mårell, A. et al. (2018). Ungulates increase forest plant species richness to the benefit of non-forest specialists. *Global Change Biology*, 24(2), e485–e495.

Bradshaw, R.H.W., Hannon, G.E., Lister, A.M. (2003). A long-term perspective on ungulate-vegetation interactions. *Forest Ecology and Management*, 181(1–2), 267–280.

Chollet, S., Baltzinger, C., Ostermann, L., Saint-André, F., Martin, J.-L. (2013). Importance for forest plant communities of refuges protecting from deer browsing. *Forest Ecology and Management*, 289(1), 470–477.

Chollet, S., Baltzinger, C., Maillard, M., Martin, J.-L. (2021). Deer exclusion unveils abiotic filtering in forest understorey plant assemblages. *Annals of Botany*, 123(3), 371–381.

Cocquelet, A., Mårell, A., Bonthoux, S., Baltzinger, C., Archaux, F. (2019). Direct and indirect effects of ungulates on forest birds' nesting failure? An experimental test with artificial nests. *Forest Ecology and Management*, 437(1), 148–155.

Davis, M.A., Grime, J.P., Thompson, K. (2000). Fluctuating resources in plant communities: A general theory of invasibility. *Journal of Ecology*, 88(3), 528–534.

Defleur, A.R. and Desclaux, E. (2019). Impact of the last interglacial climate change on ecosystems and Neanderthals behavior at Baume Moula-Guercy, Ardèche, France. *Journal of Archaeological Science*, 104(April), 114–124.

4. https://www.decadeonrestoration.org/.

Garrido, P., Mårell, A., Öckinger, E., Skarin, A., Jansson, A., Thulin, C.-G. (2019). Experimental rewilding enhances grassland functional composition and pollinator habitat use. *Journal of Applied Ecology*, 56(4), 946–955.

Gill, J.L. (2014). Ecological impacts of the late quaternary megaherbivore extinctions. *New Phytologist*, 201(4), 1163–1169.

Green, A.J., Baltzinger, C., Lovas-Kiss, Á. (2022). Plant dispersal syndromes are unreliable, especially for predicting zoochory and long-distance dispersal. *Oikos*, 2022(2), e08327 [Online]. Available at: https://doi.org/10.1111/oik.08327.

Hidding, B., Tremblay, J.-P., Côté, S.D. (2013). A large herbivore triggers alternative successional trajectories in the boreal forest. *Ecology*, 94(12), 2852–2860.

Jaroszewicz, B. (2013). Endozoochory by European bison influences the build-up of the soil seed bank in subcontinental coniferous forest. *European Journal of Forest Research*, 132(3), 445–452.

Jones, C.G., Lawton, J.H., Shachak, M. (1994). Organisms as ecosystem engineers. *Oikos*, 69(3), 373–386.

Laurent, L., Mårell, A., Korboulewsky, N., Saïd, S., Balandier, P. (2017). How does disturbance affect the intensity and importance of plant competition along resource gradients? *Forest Ecology and Management*, 391(May), 239–245.

Lortie, C.J., Brooker, R.W., Choler, P., Kikvidze, Z., Michalet, R., Pugnaire, F.I., Callaway, R.M. (2004). Rethinking plant community theory. *Oikos*, 107(2), 433–438.

Lundberg, J. and Moberg, F. (2003). Mobile link organisms and ecosystem functioning: Implications for ecosystem resilience and management. *Ecosystems*, 6(1), 87–98.

Lundgren, E.J., Ramp, D., Rowan, J., Middleton, O., Schowanek, S.D., Sanisidro, O., Carroll, S.P., Davis, M., Sandom, C.J., Svenning, J.-C. et al. (2020). Introduced herbivores restore late Pleistocene ecological functions. *Proceedings of the National Academy of Sciences*, 117(14), 7871–7878.

Mårell, A., Archaux, F., Korboulewsky, N. (2009). Floral herbivory of the wood anemone (*Anemone nemorosa* L.) by roe deer (*Capreolus capreolus* L.). *Plant Species Biology*, 24(3), 209–214.

Martin, J.-L. and Baltzinger, C. (2002). Interaction among deer browsing, hunting, and tree regeneration. *Canadian Journal of Forest Research*, 32(7), 1254–1264.

Milotić, T. and Hoffmann, M. (2016). Reduced germination success of temperate grassland seeds sown in dung: Consequences for post-dispersal seed fate. *Plant Biology*, 18(6), 1038–1047.

Mitchell, F.J.G. (2005). How open were European primeval forests? Hypothesis testing using palaeoecological data. *Journal of Ecology*, 93(1), 168–177.

Mohr, D., Cohnstaedt, L.W., Topp, W. (2005). Wild boar and red deer affect soil nutrients and soil biota in steep oak stands of the Eifel. *Soil Biology and Biochemistry*, 37(4), 693–700.

Murray, B.D., Webster, C.R., Bump, J.K. (2013). Broadening the ecological context of ungulate–ecosystem interactions: The importance of space, seasonality, and nitrogen. *Ecology*, 94(6), 1317–1326.

Nuttle, T., Ristau, T.E., Royo, A.A. (2014). Long-term biological legacies of herbivore density in a landscape-scale experiment: Forest understoreys reflect past deer density treatments for at least 20 years. *Journal of Ecology*, 102(1), 221–228.

Olmsted, C.F., Betras, T., Pasquini, S.C., Destefano, S., Faison, E., Carson, W.P. (2021). Characteristics of stem-breaking by moose (*Alces alces*, Cervidae): A case-study and worldwide review. *The Journal of the Torrey Botanical Society*, 147(4), 304–315.

Pendergast, T.H., Hanlon, S.M., Long, Z.M., Royo, A.A., Carson, W.P. (2016). The legacy of deer overabundance: Long-term delays in herbaceous understory recovery. *Canadian Journal of Forest Research*, 46(3), 362–369.

Persson, I.-L., Danell, K., Bergström, R. (2005). Different moose densities and accompanied changes in tree morphology and browse production. *Ecological Applications*, 15(4), 1296–1305.

Rico, Y., Boehmer, H.J., Wagner, H.H. (2014). Effect of rotational shepherding on demographic and genetic connectivity of calcareous grassland plants. *Conservation Biology*, 28(2), 467–477.

Royo, A.A., Stout, S.L., Decalesta, D.S., Pierson, T.G. (2010). Restoring forest herb communities through landscape-level deer herd reductions: Is recovery limited by legacy effects? *Biological Conservation*, 143(11), 2425–2434.

Saatkamp, A., Henry, F., Dutoit, T. (2021). Romans shape today's vegetation and soils: Two millennia of land-use legacy dynamics in Mediterranean grasslands. *Ecosystems*, 24(5), 1268–1280.

Stockton, S.A., Allombert, S., Gaston, A.J., Martin, J.-L. (2005). A natural experiment on the effects of high deer densities on the native flora of coastal temperate rain forests. *Biological Conservation*, 126(1), 118–128.

Svenning, J.-C. (2002). A review of natural vegetation openness in north-western Europe. *Biological Conservation*, 104(2), 133–148.

Tanentzap, A.J., Burrows, L.E., Lee, W.G., Nugent, G., Maxwell, J.M., Coomes, D.A. (2009). Landscape-level vegetation recovery from herbivory: Progress after four decades of invasive red deer control. *Journal of Applied Ecology*, 46(5), 1064–1072.

Tremblay, J.-P., Huot, J., Potvin, F. (2006). Divergent nonlinear responses of the boreal forest field layer along an experimental gradient of deer densities. *Oecologia*, 150(1), 78–88.

Vera, F.W.M. (2000). *Grazing Ecology and Forest History*. CABI Publishing, Wallingford.

Vild, O., Hédl, R., Kopecký, M., Szabó, P., Suchánková, S., Zouhar, V. (2017). The paradox of long-term ungulate impact: Increase of plant species richness in a temperate forest. *Applied Vegetation Science*, 20(2), 282–292.

Waldram, M.S., Bond, W.J., Stock, W.D. (2008). Ecological engineering by a mega-grazer: White rhino impacts on a South African Savanna. *Ecosystems*, 11(1), 101–112.

Wiegmann, S.M. and Waller, D.M. (2006). Fifty years of change in northern upland forest understories: Identity and traits of "winner" and "loser" plant species. *Biological Conservation*, 129(1), 109–123.

Wilby, A., Shachak, M., Boeken, B. (2001). Integration of ecosystem engineering and trophic effects of herbivores. *Oikos*, 92(3), 436–444.

14

A Historical Ecology of the Compiègne Forest (N France)

Jérôme BURIDANT, Boris BRASSEUR, Hélène HOREN, Emilie GALLET-MORON and Guillaume DECOCQ
Jules Verne University of Picardie, Amiens, France

14.1. Introduction

The Compiègne forest (49°22′ N; 2°54′ E; 32–148 m altitude) is the third biggest state forest in the French lowlands. It is a sub-Atlantic temperate oak-beech forest located at 70 km north-east to Paris, which covers ca. 14,500 ha at the junction between the Aisne and the Oise rivers. The substrate is dominated by acidic sands and limestones. It has long been considered as a very ancient woodland, which might have continuously existed since the beginning of the last post-glacial afforestation episode, ca. 10,000 years ago. This belief relies on the fact that (i) Historians of the 19th century erroneously interpreted Caesar's landscape description (*De Bello Gallico*) as a dense woodland at the time of the Romans conquest (1st century BC); (ii) following the Roman empire collapse (5th century AD), the forest became a royal hunting reserve until the French revolution in 1789 and (iii) since the 18th century, the forest is a state forest managed for wood production and stag hunting. However, evidence is accumulating that the Compiègne forest may not be as ancient as previously thought, and has been patterned by thousands of years of human activities. Here, we present a synthesis of recent research in historical ecology that has been conducted in this forest by the EDYSAN research group in Amiens (Jules Verne University of Picardie) to better understand the legacy of past human–forest interactions in the current forest ecosystem.

14.2. The ancient forest: an intensively managed agricultural landscape?

Landscape archaeology using aerial photographs developed in England and France from the 1950s onward (Dassié 1978). In northern France, it revealed a huge density of archaeological imprints (e.g. buried calcareous stones into silty topsoils) across agricultural landscapes, mostly of Gallo-Roman origin, indicating that the land was already intensively cultivated at that time (Agache 1978). In comparison, woodlands remained largely unexplored since the vegetation cover hides micro-reliefs, buried walls and other archaeological remains. Extensive archaeology using walking transects in the understories can help in finding visible topographic anomalies and artifacts, but may be limited by a dense ground vegetation or by management-induced disturbances.

In the case of large forests, such as the Compiègne forest, which were already present during the "forest minimum" (ca. 1850 in France; Bergès and Dupouey 2017) and have been continuously mapped over the last few centuries, it has been postulated that they likely exist since prehistoric times. This hypothesis was seemingly supported by ancient written sources. Pliny the Elder and Suetonius, for example, designed Gaul as *Gallia comata*, which means "hairy Gaul", the Gaul prior to the Roman conquest (Clavel-Lévêque 1989). This was interpreted by historians of the 19th century as evidence for a land covered by forests (Maury 1848, 1850). However, the first archaeological explorations of the 19th century started to cast doubt on the ancient forest hypothesis, since a number of Gallo-Roman sites were discovered, such as *vici* (i.e. urban areas), *villae* (i.e. isolated farms), a sanctuary and several roads.

From 1997 to 2017, an intensive field survey was implemented, which revealed more than 300 Roman sites, mostly rural settlements, within the forest (Thuillier 2017). This field survey was complemented by a remote sensing survey using the LiDAR technology (Box 14.1). A first flight covering a small part of the forest was conducted in 2011 at 9 points per m^2; a second one covered the entire forest in 2014 at a higher resolution (23 points per m^2). A total of 5,710 anomalies were evidenced, including 1,457 one-off (e.g. depressions, bomb holes), 3,797 linear (e.g. banks, roads) and 456 zonal (e.g. mounds, enclosures) anomalies (David 2014). All anomalies that cannot be identified with certainty were verified in the field. Some of them corresponded to prehistoric or protohistoric artifacts (e.g. dolmen, cutting workshops, barred spurs), but the vast majority of them dated back the Gallo-Roman period (1st–4th century AC). Only a sandy area located towards the center-east of the forest was free from occupation sites (Figure 14.1).

LiDAR has rapidly become a major tool in historical ecology. LiDAR (for light detecting and ranging) is a technique for measuring distances using the properties of laser beams. The principle is to analyze the return time of a light beam between its emission source and an object. In forest, topographic LiDAR is mainly used. It is a LiDAR carried by an aircraft flying at low or medium altitude, measuring the topography over a relatively large area. It allows us to assess the relative altitude of the ground and objects at a high precision (a few cm). The laser beam is reflected by the canopies, by tree trunks and branches and by the ground. It is therefore possible to classify the cloud of points obtained by selecting the upper points, the intermediate points and the ground points, to develop a digital elevation model (DEM), a digital terrain model (DTM) and intermediate models. DTMs have archaeological or geomorphological applications. Elevation models and intermediate models rather have dendrometric applications, to assess standing timber volumes and the structure of forest stands. The higher the number of impacts reaching the ground (in points per m^2), the higher the resolution of the resulting model. In Western Europe, the first archaeological applications of LiDAR were developed in the Rastatt forest, in Baden (Sittler and Hauger 2007). In France, the first archaeological survey was carried out in the forest of Haye, east to Nancy (Georges-Leroy et al. 2011). LiDAR is now used in a variety of applications, and several European countries have developed programs to cover the entire national territory.

Box 14.1. *The LiDAR technology*

The results show that during the Gallo-Roman period, human activities were structured by an important Roman road, connecting two cities: Soissons (*Augusta Suessionum*), to the west, and Senlis (*Augustomagus*), to the south-west. The primary road crossed three secondary towns (*vici*), and several farms (*villae*) were scattered on both sides, the more distant ones being connected to the main road by secondary roads. The LiDAR also revealed that over almost the whole forest area, the land was divided into rectangular parcels, which were also consistently structured by the ancient road network (Figure 14.2). They are very similar to those retrieved in other French forests, which have been shown to be the ancient agricultural lands, where plowing led to large geometric areas surrounded by banks possibly topped by a hedgerow (Georges-Leroy et al. 2011). In some parts of the forest, smaller parcels of more irregular shape sit on the top of these typical parcels of lands, probably of medieval origin (5th–15th century) (Buridant et al. 2020). This indicates that parts of the current forest were still cultivated in the Middle Ages.

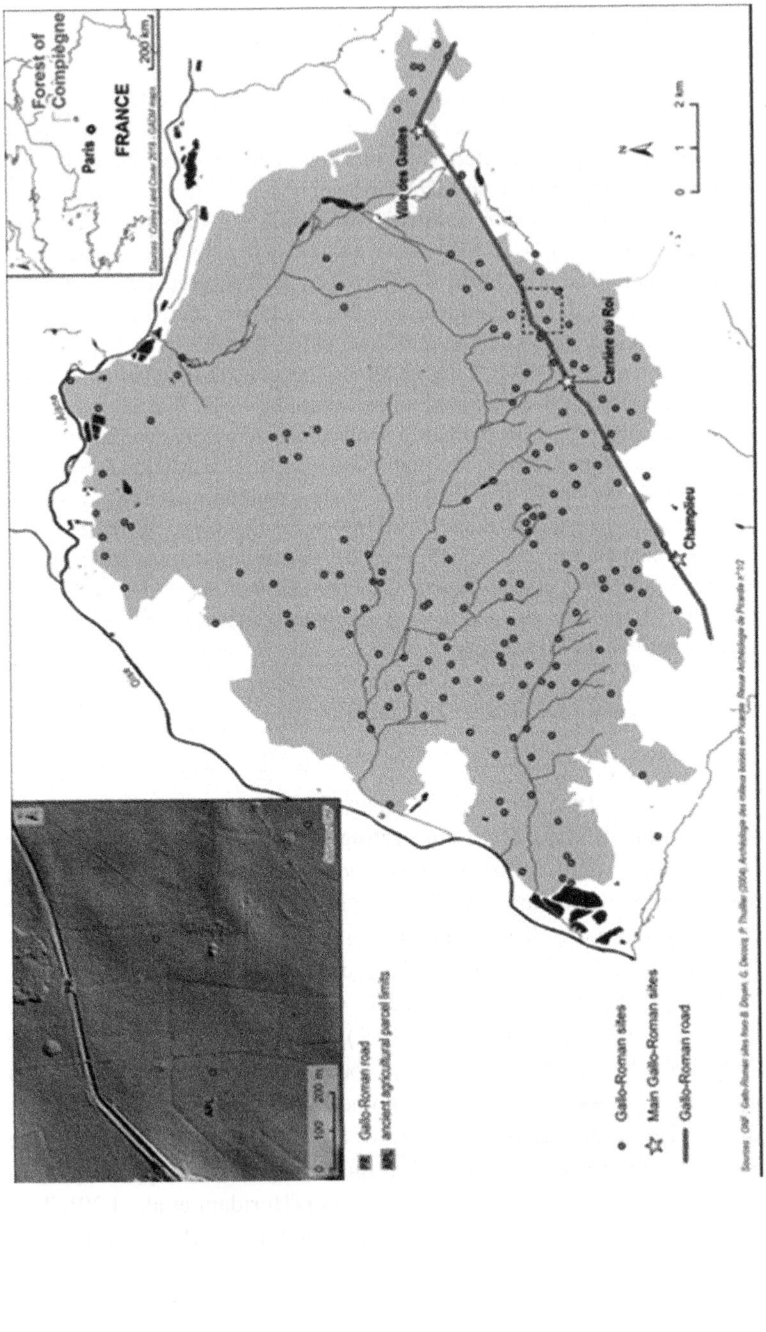

Figure 14.1. *Former Gallo-Roman settlements in the current Compiègne forest. The top-right map shows the location of the forest in Northern France. The top-left image is a digital terrain model from the LiDAR, which reveals the ancient parcels of agricultural land. For a color version of this figure, see www.iste.co.uk/decocq/ecology.zip*

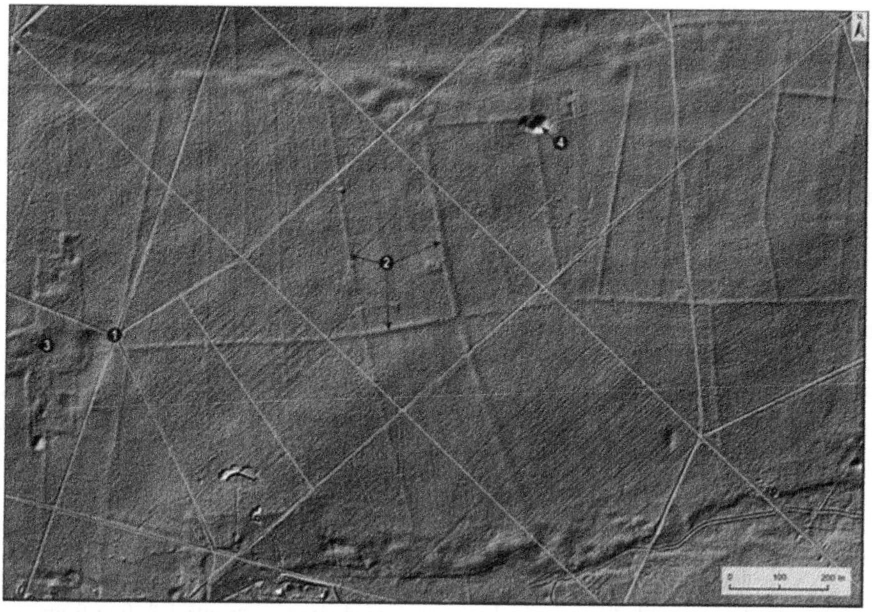

Figure 14.2. *LiDAR image of part of the Compiègne forest. The current network of star-shaped roads (1) is inherited from the 16th to 18th century (yellow lines). We can see under this network the rural territory organization from the Antiquity made up of large rectangular agricultural parcels (2), a residential area (3) and limestone extraction areas (4). For a color version of this figure, see www.iste.co.uk/decocq/ecology.zip*

An interesting feature of this "fossilized" agricultural landscape revealed by the LiDAR-derived digital elevation model is the presence of many closed circular depressions (CCD) that are systematically located along an inner edge of the rectangular parcels (Figure 14.3). They consist of 0.5 to 3 m-deep depressions with a diameter ranging from 20 to 60 m. A pedo-sedimentary analysis of cores taken outside, on the edge and at the center of 10 CCD, was conducted to determine their origin. The shallowest CCD (sCCD) revealed a single-phase, non-stratified sediment infill. The filling material was clearly anthropogenic, consisting of a mixture of local soils, their limestone substrate and sometimes fragments of ceramics. At the top, soil horizon development occurred. In contrast, the deepest CCD (dCCD) showed no or little infilling and the limestone substrate was encountered at <50 cm. Both CCD types had a bathtub shape with a flat bottom at the interface with the geological substrate, which is typical of an anthropogenic excavation. However, CCD could not be limestone quarries used to provide the building material – that are also present elsewhere in the landscape – since the hard limestone (e5b) was intact. The chemistry of the local substrate also excluded a karstic (sinkhole) or salt diapir dissolution origin (Delafosse 1948; Etienne et al. 2011). These CCD thus correspond to ancient pits, where limestone powder and calcareous marls and loess (e5b/c) were extracted. As we found no accumulation or overlying of these materials around CCD, we hypothesize that they were sprayed over agricultural lands to buffer the natural soil acidification under a humid temperate climate, thereby maintaining productivity. Liming was indeed a common practice for Gauls and Gallo-Roman, as mentioned by Latin agronomists (Varro, Pliny the Elder, Columella). The abundance of CCD (>100) in the Compiègne forest indicates that liming was a widespread practice during the Roman times. It not only creates an artificial topography and localized thick anthropogenic soils but also likely induced a general alkalinisation of soils, which durably impacted soil properties in former fields. Recent work suggests that liming effects on soil pH last several millennia (Brasseur et al. 2018), and hence can still impact vegetation and ecosystem functioning (Box 14.2).

To test whether past Gallo-Roman settlement still affect the present characteristics of soils, vegetation and soil seed bank, we designed a comparative study (Plue et al. 2008). To each of 24 former isolated Gallo-Roman *villae* (i.e. "sites", where an inhabited building used to be settled), we paired an unoccupied site (i.e. "controls", free from any former human settlement but not from past human influence) at a distance of <200 m, which shared the same topography, substrate and management history. An 800 m^2 quadrat was disposed in each site and control, into which all vascular plant species were recorded, and soil samples were taken for further soil chemical analyses and permanent seed bank assessment (25 cores per quadrants, 26 cm diameter, 20 cm depth after litter removal).

The soil was persistently altered on the sites, resulting in elevated phosphorus levels and pH (dependent on initial soil conditions). The upper soil layers also contained scattered stone fragments, pieces of limestone and shards of roof tiles (tegulae). These altered soil features translated into significantly increased vegetation cover and seed bank density (1,500 seeds m^{-2} vs. 333 seeds m^{-2}), increased species richness in both vegetation (37.1 vs. 23.2) and soil seed bank (6 vs. 4) and increased compositional similarity among sites (Raup & Crick index: 0.53 vs. 0.76), as compared to unoccupied sites. In particular, the vegetation on Gallo-Roman sites had a significantly more competitive and ruderal but less stress-tolerant character than controls. Nitrogen-demanding and calciphileous species were strongly associated with sites, while acidiphilous species were significantly associated with controls. It is noteworthy that fruit trees (e.g. apple tree, medlar tree) often appeared in the woody layers on the archaeological site or nearby. The local persistence of ruderal species below forest canopies is surprising. It is likely facilitated by forest management, which allows periodic light arrival at the forest floor, hence their recruitment from the soil seed bank and the replenishment of the latter.

These results revealed that 1,600 years after their reclamation and forest colonization, Gallo-Roman settlements still leave a persistent trace on forest soil, vegetation and seed bank. By creating entirely different habitats with irreversibly (?) altered geochemical features, these archaeological sites form plant species-rich islands in an otherwise species-poor forest ocean, thereby contributing to biodiversity at the entire forest scale (Closset-Kopp and Decocq 2015). Furthermore, the Roman legacy in vegetation is not limited to former residences. Roman roads have likely served as important corridors for species, before and after their reclamation. For example, in the Compiègne forest, the archaeophyte *Vinca minor* has been observed only on the former Roman road.

Box 14.2. *Impacts of Roman settlements on biodiversity and ecosystem functioning*

The area of the present-day forest of Compiègne was thus previously not a forest, but rather looked like an open agricultural land. Archaeological studies and pollen analyses suggest that the local deforestation started long before the Roman conquest (Perrière and Leroyer 2006). The first populations of farmers settled around 4500 BC in the Oise and Aisne valleys, which border the current Compiègne forest at the west and north, respectively (Figure 14.1). At that time, the pollen diagrams are dominated by oak (*Quercus*) and hazelnut (*Corylus*), with lime (*Tilia*), elm (*Ulmus*) and alder (*Alnus*). This typical vegetation of the Atlantic period (5500–2300 BC) was probably an open-canopy forest. Beech (*Fagus sylvatica*) first appeared after 3000 BC and became common only after 1250 BC (Leroyer et al. 2017).

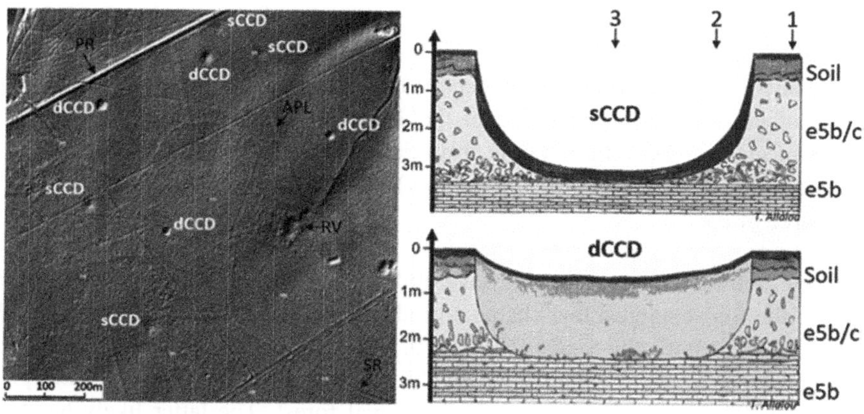

Figure 14.3. *Left: LiDAR-derived digital elevation model revealing closed circular depressions filled (sCCD) or not (dCCD) by sediments. Other archaeological features are visible: a Gallo-Roman villa (RV), Senlis-Soissons primary road (PR), a secondary road (SR) and agricultural parcel limit (APL). Right: Geological profile of a dCCD (top) and sCCD (bottom). e5b/c Eocene (upper Lutetian) powdery limestone, e5b Eocene (middle Lutetian) hard limestone. 1, 2 and 3 show the core position for each of the CCD. For a color version of this figure, see www.iste.co.uk/decocq/ecology.zip*

14.3. The Medieval forest: a woodland (re)birth or a savanna-like ecosystem?

Gallo-Roman settlements were quite suddenly abandoned towards the end of the 4th century AD, with archaeological evidence for destruction and fires as soon as the 3rd century AD (Frémont and Woimant 1975, 1976; Tuffreau-Libre 1977). A woodland is mentioned for the first time at the 6th century by Grégoire de Tours (Histoire des Francs), when he relates the death of the Merovingian king Clotaire I in 561, from a fever "while he was hunting in the forest of Cuise" (*dum in Cotia silva venationem exercet*). The forest of Cuise is presented as a hunting forest belonging to the royal domain. Since this first mention, historical texts confirm the continuous presence of a woodland in this area, which was later renamed Compiègne forest.

The forest colonization of abandoned fields occurred in the course of secondary succession, so that the early Medieval forest may have consisted of relatively dense oak-dominated woodlands. This hypothesis is supported by the low human population density in this area, following the Great Invasions, and the fact that the wood framework of local houses was built using tall trees with very narrow tree rings, suggesting they grew in packed populations, a typical characteristic of early-successional forest communities. The human pressure on the forest likely increased again with the demographic booming of the 12–13th century. Archaeological studies evidenced activities associated with the supply of wood, such as pottery workshops (starting at the 10–11th century), glass factories (12th century) and metalworking industries (14th century), all established into the forest. The number of glass factories increased in the 14th and 15th centuries, when several families of glass-makers immigrating from Eastern France established there. At this time, the bracken fern (*Pteridium aquilinum*) was likely abundant, since it was the only source of lye.

Dating of buried charcoals retrieved from kilns in topographic depressions in the center of the forest suggest that charcoal production was conducted at least from the 12–13th century until the years 1610–1670. However, no written source specifies the structure, extent and composition of the forest before the 16–17th century, so that it is challenging to reconstruct the Medieval forest. The latter likely has been patterned by two main drivers: user rights and hunting.

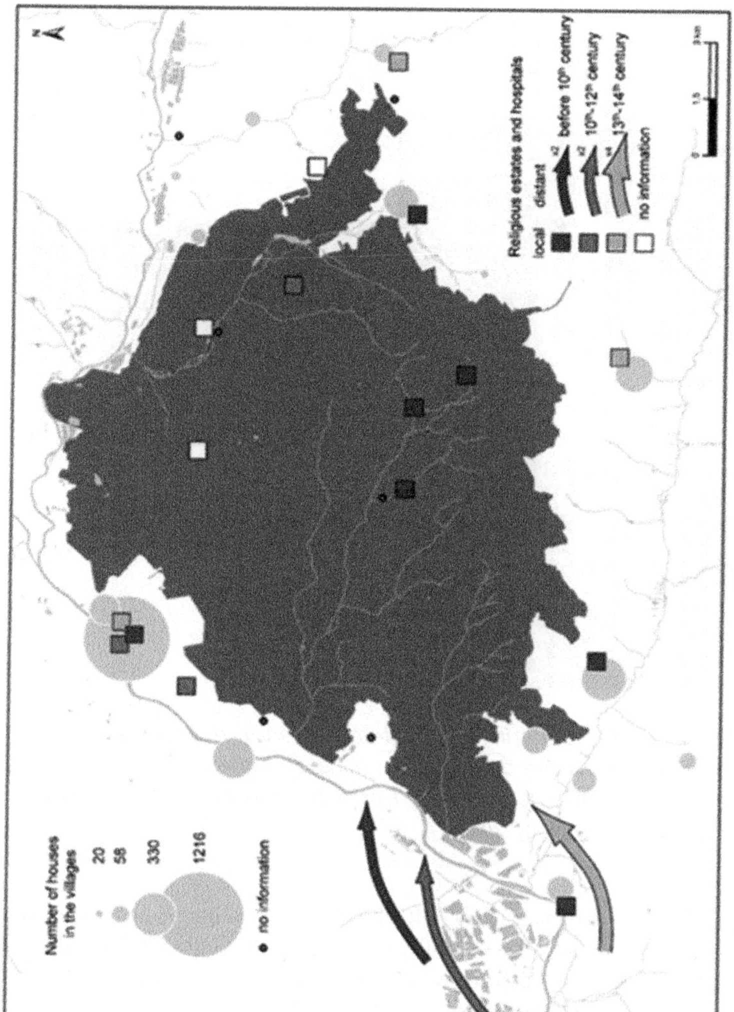

Figure 14.4. *State of play for user rights in the Compiègne forest in 1663. For a color version of this figure, see www.iste.co.uk/decocq/ecology.zip*

Pedological pits were made in the center of the forest, close to a dune-like relief free from any archaeological artifact as revealed by the LiDAR imagery. The pits showed podzols covered a more or less thick layer of drift sands. The abundance of dung beetle galleries observed in one horizon of the profile (see Figure 14.5 below) suggests locally abundant animal waste associated with a pasture land-use, and thus a rather open landscape. Consistently, the sand layer observed on the top of the paleosol may reflect intensive grazing-induced drift sands on sandy soils.

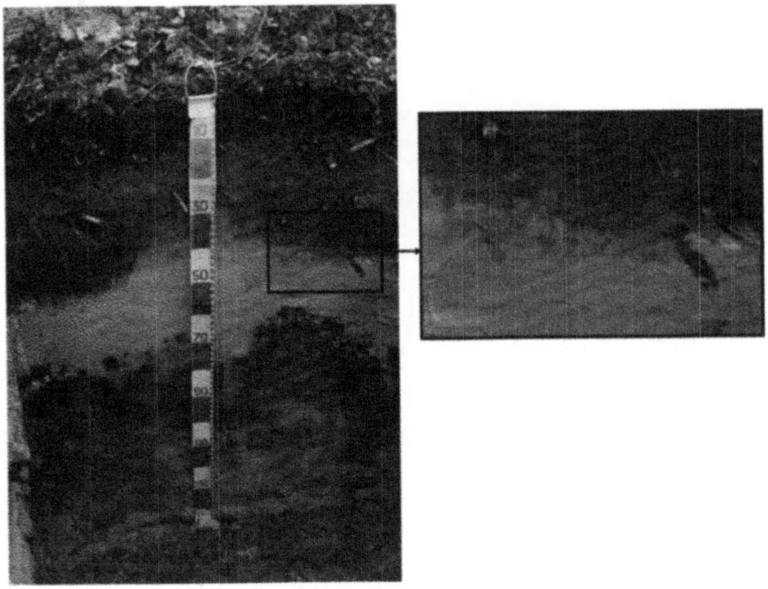

Figure 14.5. *Podzol profile in the drift-sands area (left panel). Close up of the contact between the dark organo-mineral horizon and the ash grey eluvial horizon just below (right panel). For a color version of this figure, see www.iste.co.uk/decocq/ecology.zip*

Podzol profile in the drift sands area (left panel). The soil is buried under ca. 35 cm of sand. The abundance of dung beetle galleries observed at contact between the dark organo-mineral horizon and the ash gray eluvial horizon just below (right panel) suggests local soil enrichment associated to a pasture land-use.

To obtain a picture of past local species assemblages, we sampled soil charcoals in a podzol and two others soil types within currently mature forest stands (Feiss et al. 2017b). The radiocarbon dating returned no charcoals older than the 10–11th century. We found no relationship between the charcoal age and their vertical distribution, due to soil mixing associated with belowground biological activity. Soil charcoal assemblages thus hardly reflect temporal changes in forest stand composition. However, the fact that the oldest charcoals came from birch (*Betula*) supports the hypothesis that the 10–11th-century forest was very open, with a substantial proportion of shade-intolerant tree species. Light-demanding, early-successional taxa, while missing from current tree communities, were almost systematically more abundant than shade-tolerant species across the three sites. This suggested that the forest canopy used to be much more open than today, consistent with a savanna-like landscape maintained by the intense human pressure and heavy grazing by wild and domestic animals (Feiss et al. 2017a).

Box 14.3. *What do soil archives teach us?*

Like many European forests in the Middle Ages, the Compiègne forest was not exclusively intended for wood production. Many uses were practiced by the local populations, such as picking up dead wood, cutting green wood and grazing by cattle and pigs. From the 13 to 14th century onward, these uses became user rights granted to local populations and institutions. These user rights were progressively restricted, especially by the forest reformation of 1663, to limit the overexploitation of the forest resources (Figure 14.4). Still, such rights were conceded to 16 communities of inhabitants, 25 ecclesiastical communities and one glass furnace. Some communities have their rights restricted to certain areas, while many others can take their herds throughout the entire forest. The authorized load often exceeded one animal per hectare, a density much greater than the carrying capacity of the forest. It is thus likely that the forest landscape at this time rather looked like a wooded pasture (i.e. a temperate savanna; Rackham 1998) than a closed-canopy woodland, a hypothesis supported by soil charcoal analyses (Box 14.3).

At the same time (from the 17th century onward), with the progress in firearms, pheasant hunting became a fashion. Kings Louis XIV and Louis XV built seven pheasantries in the center of the forest. Pheasantries are closed, cultivated parks dedicated to the breeding of pheasants, which were then released in large open spaces (called "tirés") into the forest, allowing them to take off. This practice was still ongoing in the 19th century, since Emperor Napoleon III was a lover of pheasant hunting and was often coming in the Compiègne forest for this purpose.

However, the Compiègne forest was also dedicated to the production of wood. In the 17th and 18th centuries, logs were floated to Paris via the Oise river, where they were used for construction and as fuel. Interestingly, the forest provided a lot of short, curved logs to royal shipyards on the Atlantic coast, as well as to a local shipyard, which built barges for inland water shipping. The abundance of short, curved logs suggests that the core of the forest, which was managed as a high forest for hunting, was surrounded by more intensively managed short-rotation, dense coppice-like woodlands.

14.4. The contemporary forest (19th century onward): a closed-canopy multifunctional woodland

Today's forest landscape is mostly a legacy of recent periods of restoration and replanting, which started in the mid-18th century (Figure 14.6). From 1772 to 1792, deciduous trees, mainly pedunculate oaks, were planted by Pierre-Lucien Pannelier d'Annel and his son, Antoine-Lucien Pannelier. A second phase of reforestation occurred throughout the 19th century, over 6,610 ha (i.e. 46% of the total forest area) (Buridant 2008). Following World War II, conifers, mainly Scots pine, were widely planted over the less fertile areas. More than half of the forest is therefore composed of recent plantations (<250 years old). It is likely that human disturbances associated with these plantations have facilitated the invasive spread of the American black cherry (*Prunus serotina*), which started towards the mid-19th century. The Compiègne forest is currently the most invaded in France, with ca. 80% of the forest area where the invader occurs, as isolated trees to monospecific stands, thereby significantly impacting the recipient ecosystem (Box 14.4).

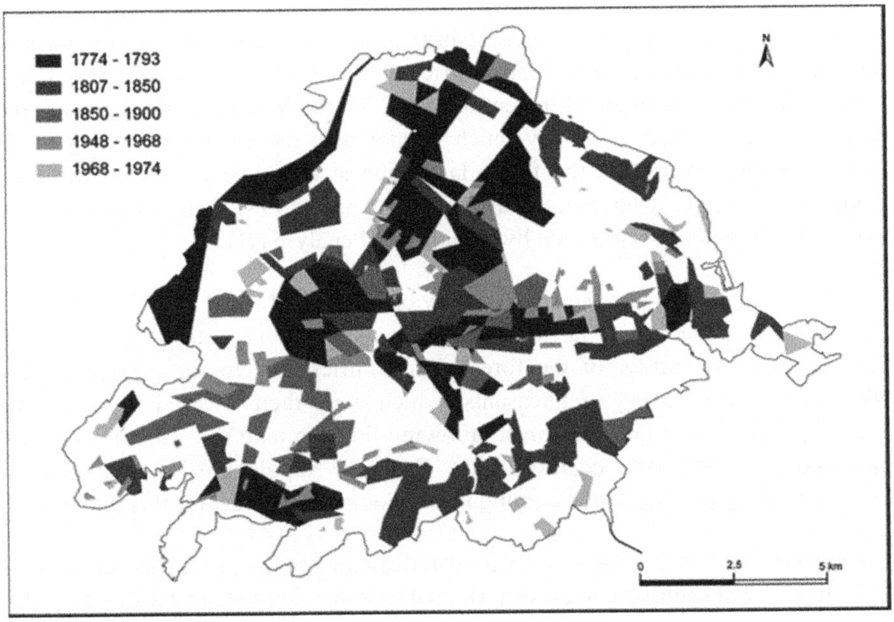

Figure 14.6. *Tree plantation in the Compiègne forest from 1774 to 1974. For a color version of this figure, see www.iste.co.uk/decocq/ecology.zip*

The current forest is mostly dedicated to timber production (oak, beech and Scots pine) and stag-hunting. As a public sub-urban forest, the forest also experienced an increasing pressure from local populations for their recreative activities. Two nature reserves have been created, one in the 19th century (Beaux Monts) and the other in the 1970s (Grands Monts). These reserves are useful tools to analyze the impact of the current forest management on vegetation (Box 14.5).

The first decades of the 21st century are marked by rapid changes in the forest vegetation in response to global changes (Box 14.5). These changes are event accelerating: most canopy tree species experienced severe dieback due to several non-independent factors: repeated drought episodes (beech), cockchafer (oak) and bark beetle (spruce) boom, chalararosis epidemics (ash), etc. As a result, light arrival at the forest floor increases with dramatic changes in the ground vegetation to come.

Prunus serotina either escaped from the castle park or has been introduced directly in the Compiègne forest around 1850 but has been reported as an invader only in 1970. It has now spread in the entire forest. Many studies have been conducted to assess the impacts of the invasion, by comparing tree physiology, plant communities, soil properties and disturbance history between *P. serotina*-invaded and uninvaded paired stands. Overall, species assemblages in invaded and uninvaded stands were similar (Chabrerie et al. 2008). Habitat conditions and disturbances explained most of the variation in both plant diversity and *P. serotina* density, the last two factors being weakly associated. The herb layer of invaded stands exhibited significantly more specialist species and a lower trait diversity compared to uninvaded stands. Traits characterizing shade-tolerant, short-living ruderals and shade-avoiders (vernal geophytes) were significantly associated with invaded stands, while those associated with light-demanding graminoids characterized uninvaded stands. This indicates that *P. serotina* was becoming the new ecosystem engineer, first by inducing trait convergence and community specialization, thus promoting traits that enable species to capture resources in the new environment it was creating, and second by reducing the grain of local heterogeneities (Chabrerie et al. 2010). Compared to indigenous tree species, P. serotina exhibited higher foliar N, foliar P and lower foliar C:N and N:P ratios (Aerts et al. 2017). *P. serotina* also affected foliar nutrient contents of co-occurring indigenous tree species leading to decreased foliar N and increased C:N ratio in beech, decreased foliar N:P ratio in hornbeam and beech and increased foliar P in Scots pine in invaded versus uninvaded stands. *P. serotina* thus changes nitrogen, phosphorus and carbon cycles to its own advantage, hereby increasing carbon turnover via labile litter, affecting the relative nutrient contents in the overstory leaves and potentially altering the photosynthetic capacity of the other broadleaved species.

Box 14.4. *Impacts of* Prunus serotina *invasion on forest vegetation*

Recent vegetation changes have been recently documented using a resurvey study and a life-history trait-based approach (Closset-Kopp et al. 2019). This study highlighted decadal changes (1970–2015) in the ground vegetation and searched for the drivers behind these changes. A number of process-based hypotheses were tested using paired comparisons (old vs. new records) of community-weighted means and conditional inference classification trees for a number of traits across soil and management types. Compared to unmanaged stands (nature reserves), managed stands underwent the most pronounced vegetation changes, as a likely effect of modifications in forest harvesting practices since 1970. Consistent with general trends observed in European forests, species richness of vascular plants increased at both stand (α-diversity) and forest (γ-diversity) scales, due to the non-random, directional colonization by the same suite of species, causing compositional, functional and phylogenetic homogenization among habitats (i.e. decreased β-diversity). Forest management, via the repeated passing of heavy forestry vehicles, emerged as the key driver of these changes. By altering canopy structure, it increased not only light availability at the forest floor but also the vulnerability of understories to climate warming and atmospheric deposits. Furthermore, increasingly heavy forestry vehicles that drive more frequently across forest stands likely generate microhabitats suitable for ferns, graminoids and N-demanding forbs and also act as dispersal agents.

Box 14.5. *Vegetation changes at the Anthropocene*

14.5. Conclusion

Altogether, these results demonstrate that the Compiègne forest has established on an ancient cultivated plain and on neighboring plateaus, where many people used to live, work and trade for centuries. In this context, little place was left for woodlands during the Gallo-Roman period. The rural landscape rather looked like an intensively managed open field, maybe with a few small woodland patches restricted to the steepest slopes of surrounding hills and central sand dunes. The forest established progressively during the early Middle Ages, from the spontaneous afforestation of abandoned farmlands. Since the secondary succession was strongly constrained by human activities, wild and domestic herbivores, it is likely that the vegetation rapidly shifted from a dense early-successional woodland to an open wooded pasture, so that the late Medieval forest likely resembled more a temperate savanna than a closed-canopy woodland. Hunting activities have been a major driver of the forest landscape through medieval, modern and contemporary times, leading to the long-lasting maintenance of open habitats within a high forest. Canopy closure and the subsequent development of a shade-tolerant flora is probably a recent phenomenon, which is currently challenged by global warming-associated environmental changes since the forest canopy tends to re-open. Most importantly, we showed that past land-use and historical human activities strongly shaped soil and vegetation features, thereby influencing today's ecosystem functioning. Current biodiversity patterns are thus a composite legacy of each of the historical periods experienced by the forest, which cannot be understood without a historical ecology perspective.

14.6. References

Aerts, R., Ewald, M., Nicolas, M., Piat, J., Skowronek, S., Lenoir, J., Hattab, T., Garzón-López, C.X., Feilhauer, H., Schmidtlein, S. et al. (2017). Invasion by the alien tree Prunus serotina alters ecosystem functions in a temperate deciduous forest. *Front. Plant Sci.*, 8, 179.

Agache, R. (1978). *La Somme pré-romaine et romaine*. Société des Antiquaires de Picardie, Amiens.

Bergès, L. and Dupouey, J.L. (2017). Écologie historique et ancienneté de l'état boisé : concepts, avancées et perspectives de la recherche. *Rev. For. Fr.*, 69, 297–317.

Brasseur, B., Spicher, F., Lenoir, J., Gallet-Moron, E., Buridant, J., Horen, H. (2018). What deep-soil profiles can teach us on deep-time pH dynamics after land use change? *Land Degr. Dev.*, 29, 2951–2961.

Buridant, J. (2008). Environnements forestiers, économie et sylviculture dans la France septentrionale, XVIIe–XIXe siècles. HdR Thesis, Université de Paris IV-Sorbonne, Paris.

Buridant, J., Pichard, C., Gallet-Moron, E. (2020). L'utilisation de la technologie lidar à la connaissance archéologique et géohistorique : exemples français. In *Temi et metodi della ricerca storica et filologica, nuove riflessioni tra Est ed Ovest*, Şipoş, S., Cepraga, D.O., Ardelean, L., Cosma, I. (eds). Transylvanian Review, vol. XXIX, suppl. no. 1.

Chabrerie, O., Verheyen, K., Decocq, G. (2008). Disentangling relationships between habitat conditions, disturbance history, plant diversity and American Black cherry (*Prunus serotina* Ehrh.) invasion in a European temperate forest. *Div. Distrib.*, 14, 204–212.

Chabrerie, O., Loinard, J., Perrin, S., Saguez, R., Decocq, G. (2010). Relating understorey plant species traits to environmental changes in forest stands heavily invaded by *Prunus serotina*. *Biol. Inv.*, 12, 1891–1907.

Clavel-Lévêque, M. (1989). La forêt gauloise vue des textes. *Puzzle gaulois, les Gaules en mémoire*. Presses Universitaires de Franche-Comté, Besançon.

Closset-Kopp, D. and Decocq, G. (2015). Remnant artificial habitats as biodiversity islets into forest oceans. *Ecosystems*, 18, 507–519.

Closset-Kopp, D., Hattab, T., Decocq, G. (2019). Do drivers of forestry vehicles also drive herb layer changes (1970–2015) in a temperate forest with contrasting habitat and management conditions? *J. Ecol.*, 107(3), 1439–1456.

Dassié, J. (1978). *Manuel d'archéologie aérienne*. Technip, Paris.

David, S. (2014). Rapport d'analyse et d'interprétation de données LiDAR ; forêts domaniales de Compiègne et Laigue (Oise). Report, Office National des Forêts, Paris.

Delafosse, W. (1948). De l'origine des mardelles de Lorraine. *Mémoire de l'Académie nationale de Metz*, 11, 63–85.

Etienne, D., Ruffaldi, P., Goepp, S., Ritz, F., Georges-Leroy, M., Pollier, B., Dambrine, E. (2011). The origin of closed depressions in Northeastern France: A new assessment. *Geomorphology*, 126, 121–131.

Feiss, T., Horen, H., Brasseur, B., Buridant, J., Gallet-Moron, E., Decocq, G. (2017a). Historical ecology of lowland forests: Does pedoanthracology support historical and archaeological data? *Quat. Int.*, 457, 99–112.

Feiss, T., Horen, H., Brasseur, B., Lenoir, J., Buridant, J., Decocq, G. (2017b). Optimal sampling design and minimal effort for soil charcoal analyses considering the soil type and forest history. *Veg. Hist. Archaeobot.*, 26, 627–637.

Frémont, J.M. and Woimant, B. (1975). Le vicus gallo-romain de la Carrière du Roi en forêt de Compiègne (Oise), 1e partie. *Revue archéologique de l'Oise*, 6, 44–46.

Frémont, J.M. and Woimant, B. (1976). Le vicus gallo-romain de la Carrière du Roi en forêt de Compiègne (Oise), 2e partie. *Revue archéologique de l'Oise*, 8, 45–49.

Georges-Leroy, M., Bock, J., Dambrine, E., Dupouez, J.L. (2011). Apport du lidar à la connaissance de l'histoire de l'occupation du sol en forêt de Haye. *Archéosciences*, 35, 117–129.

Grégoire de Tours (1836). Historiae ecclesiasticae Francorum. In *Histoire ecclésiastique des Francs, livre IV*, Guadet, J. (ed.). Société de l'histoire de France, Paris.

Leroyer, C., Barracand, G., Coussot, C., Werthe, E., Hulin, G., Jude, F., Leduc, C., Allenet de Ribemont, G. (2017). Nouvelles données sur le paysage végétal de la moyenne vallée de l'Oise au subboréal : l'étude de la séquence organique de Thourotte (Oise, France). *Quaternary*, 28(4), 471–489.

Maury, A. (1848). Recherches historiques et géographiques sur les grandes forêts de la Gaule et de l'ancienne France. In *Mémoire de la Société des antiquaires de France*, Duverger, E. (ed). Paris.

Maury, A. (1850). *Histoire des grandes forêts de la Gaule et de l'ancienne France*. A. Leleux, Paris.

Perrière, J. and Leroyer, C. (2006). Impact sur le milieu végétal et activités agropastorales des groupes néolithiques d'après la séquence pollinique d'Armancourt (60). In *Impacts interculturels au Néolithique moyen. Du terroir au territoire : sociétés et espaces*. Duhamel, P. (ed). Revue Archéologique de l'Est, suppl. 25, Dijon.

Plue, J., Hermy, M., Verheyen, K., Thuillier, P., Saguez, R., Decocq, G. (2008). Persistent changes in forest vegetation and seed bank 1,600 years after human occupation. *Landsc. Ecol.*, 23, 673–688.

Rackham, O. (1998). Savanna in Europe. In *The Ecological History of European Forests*, Kirby, K.J. and Watkins, C. (eds). CABI Publishing, Wallingford.

Sittler, B. and Hauger, K. (2007). Les apports du laser aéroporté à la documentation de parcellaires anciens fossilisés par la forêt : l'exemple des champs bombés de Rastatt en Pays de Bade. In *La mémoire des forêts*, Dupouey, J.L., Dambrine, E., Dardignac, C., Georges-Leroy, M. (eds). ONF – INRA – DRAC Lorraine, Nancy.

Thuillier, P. (2017). Dynamique des paysages de l'Antiquité au Moyen Age. La naissance de la forêt de Compiègne (Ier s. av. J.-C. – XIIIe s. ap. J.-C.). PhD Thesis, Université de Picardie Jules Verne, Amiens.

Tuffreau-Libre, M. (1977). La céramique commune gallo-romaine de la forêt de Compiègne (Oise) au Musée des Antiquités nationales. *Cah. Archéol. Pic.*, 4, 125–160.

15

The Chestnut Orchards in the Bolognese Apennines: A Vanishing Socio-ecological Habitat

Giovanna PEZZI[1], Fabrizio FERRETTI[2], Alberto MALTONI[3], Patrik KREBS[4], Marco CONEDERA[4] and Giorgio MARESI[5]

[1] *University of Bologna, Italy*

[2] *CREA Research Centre for Forestry and Wood, Arezzo, Italy*

[3] *University of Florence, Italy*

[4] *Swiss Federal Research Institute WSL, Cadenazzo, Switzerland*

[5] *Fondazione Edmund Mach, San Michele all'Adige, Italy*

15.1. Introduction

Castanea sativa Mill. (sweet chestnut) is a characteristic and iconic tree of the southern European mountain landscape. Its presence has always been tightly linked to human activities and the rural populations who cultivated the chestnut adopting different management options (orchards, high forests and coppices). The first evidence of a large-scale spread and cultivation of the species dates back to the Romans, who were primarily interested in coppice management for wood (poles) production (Conedera et al. 2004). Chestnut management for fruit production increased in the post-Roman Era and in the Middle Ages, in particular when the chestnut tree became the "bread tree" (*albero del pane*) in many mountain regions of Europe, giving rise to what some people refer to as the "chestnut civilization" (Gabrielli 1994). Existing chestnut groves still reflect the historical legacy of the

Historical Ecology,
coordinated by Guillaume DECOCQ.

past importance of the tree: for instance, the high number of selected fruit varieties (varying in ecological needs or ripening period), including nuts for flour production as staple food or high-quality fruits (*marroni* and *marroni*-like) for seasonal markets. Moreover, many phyto-toponyms related to this species and its cultivation testify its presence throughout southern Europe (Squatriti 2013).

Figure 15.1. *The study area and a view of an iconic traditional chestnut orchard. For a color version of this figure, see www.iste.co.uk/decocq/ecology.zip*

First signs of decline of the Middle Age chestnut civilization occurred in the 18th century as a result of the Little Ice Age climatic cooling, which caused widespread frost damage to chestnut trees at the most exposed sites. In the following centuries, other driving factors such as the improvement of agricultural cultivation techniques and the introduction of new crops (e.g. maize and potatoes), the onset of new diseases and the depopulation of the mountain countryside owing to industrialization caused a progressive and generalized decline in traditional chestnut cultivation (Conedera and Krebs 2008). In Italy, for instance, chestnut stands devoted to fruit production (orchards) have decreased by about 90% since the beginning of the 20th century, passing from 608,000 ha (Vigiani 1908) to 60,000 ha (Bounous 2014). Similarly, coppice stands underwent either an extension of the traditional rotation time or total abandonment.

In this chapter, we focus on chestnut orchards of the Bolognese Apennines (Italy, Figure 15.1), an iconic and ancient anthropogenic south-European landscape at risk of vanishing (Squatriti 2013). In particular, we will apply a historical ecology approach in order to analyze the past 80 years of its cultivation and propose possible future management options in order to perpetuate this unique landscape feature also in the future.

15.2. The traditional chestnut orchards

Traditional chestnut orchards (or groves) are open stands populated by grafted *C. sativa* trees for fruit production. They are multipurpose systems (Mariotti et al. 2019) with a semi-natural herb layer usually devoted to pasture with domestic animals (silvo-pastural systems, Figure 15.1). The presence of veteran trees (sporadic in other forest formations) contributes to increase the conservation importance of the habitat, as they preserve local varieties and provide favorable living conditions (e.g. trunks with holes, dead wood, large crowns) for other valuable taxa (animals, plants, lichens, mosses), including red list species (Krebs et al. 2008; Pezzi et al. 2020a).

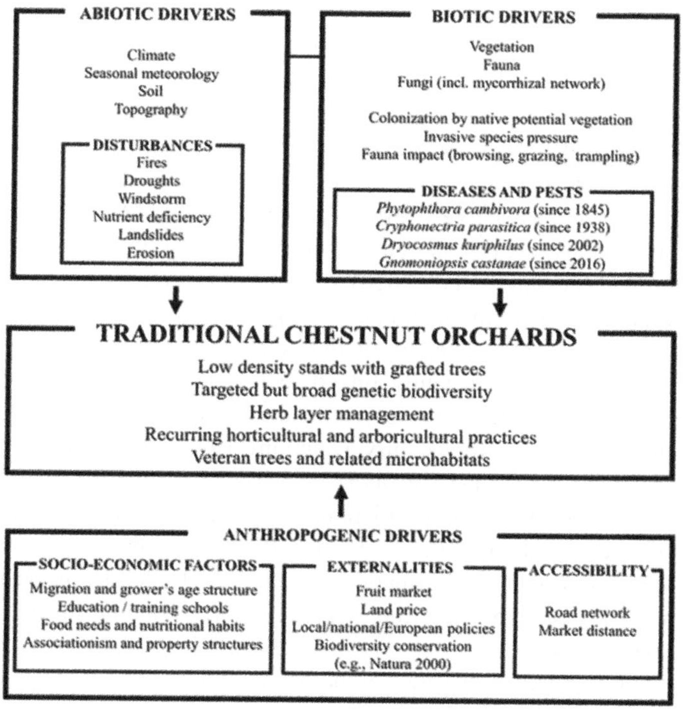

Figure 15.2. *Conceptual framework of factors affecting the chestnut orchard cultivation*

The open structure and the simultaneous presence of woody elements and open space herb species represent an ideal habitat for many plant and animal species (e.g. Obrist et al. 2011; Morelli et al. 2019). As a result, traditional chestnut orchards are biodiversity hotspots and habitats worthy of conservation (9,260 *Castanea sativa* woods) according to the European Habitat Directive (92/43/EEC). The habitat value is however the result of historical and socio-economic factors (e.g. ownership, food needs) and externalities (e.g. conservation policies, tourism, fruit market) and may thus vary highly in terms of the chestnut tree features (age, density, health, turnover, pruning, variety composition), understory vegetation and related management (mainly grazing and mowing), as well as ecological factors such as stand conditions (climate, soil, topography, accessibility), pest and pathogen incidence and other biotic interactions (e.g. game impact, propagule pressure of potential vegetation) (Figure 15.2).

15.3. The chestnut groves of the Bolognese Apennines

The Bolognese Apennines are located to the southwest of the city of Bologna (Lat. 44.492475; Long. 11.34297). They cover about 2,110 km^2 and are characterized by a marked altitudinal gradient from 49 to 1,944 m a.s.l. Chestnut cultivation in this area has been rooted since the Middle Ages. On the basis of local tradition, some authors even put forward the hypothesis that the chestnut-growing tradition in this area was actively promoted by Countess Matilda of Canossa (1046-1115), as demonstrated by the custom of designating old traditional chestnut groves as *Matildici* (Alessandri et al. 2020). Chestnuts thus represented a vital staple food for entire generations of mountain people and a commercial luxury product for the nobility and the rich city dwellers of the area. As a result, the abundance of chestnut orchards has been constantly linked to the demographic evolution, with expansion during phases of economic growth and decrease in periods of depression, such as the plague in the 14th century (Zagnoni 1997). The existence of both chestnut (i.e. *Pastonese* cultivar) and *marroni* varieties is confirmed in historical documents such as the chorographic dictionary by Calindri (1781–1783) and the coeval Boncompagni cadaster (Pezzi et al. 2020b). At the beginning of the 20th century, chestnut-dominated formations covered ca. 12,863 ha of the area, at an altitudinal range from 300 to 900 (1,000) m a.s.l., corresponding to the deciduous *Quercus*-dominated and the *Fagus sylvatica* forest belts. Most of them (i.e. 10,928 ha) were classified as chestnut orchards, as reported in 1936 by the Italian Kingdom Forest Map. This map is a very important historical reference allowing us to estimate the dramatic decline of the chestnut orchards (at present covering 1,498 ha; Regione Emilia-Romagna 2020) and their conversion into coppices or high forests over the last 85 years (Ferretti et al. 2018).

15.4. A changing world: abandonment, diseases and other problems

Similarly to other south European areas, the strategic role of chestnut culture was subjected to a first serious threat with the appearance of the ink disease (*Phytophthora cambivora* (Petri) Buism and Petri 1917) in the 19th century. Existing concerns about this disease were confirmed by Quattrocchi (1938), who found 1,024 ha of chestnut orchards affected by the ink disease in the Bologna Apennines, together with 703 ha completely degraded and 7,862 ha in bad condition because of the lack of appropriate management.

Following World War II, most Apennine villages witnessed a large migratory movement towards cities and plains. The need to find a better way of life and to escape from poverty and famine induced many young people to leave the mountains. This wave of emigration, combined with the heavy war casualties, severely affected the local communities in terms of both number of people and their links to the territory and traditions (Gallingani 2019). The resulting social change produced a progressive abandonment of chestnut culture and related activities such as tree pruning, lawn mowing, collecting foliage and burs. Moreover, the contemporary appearance and spread of the chestnut blight (*Cryphonectria parasitica* (Murr.) Barr; Biraghi 1946) affected the trees' health and fruit production, discouraging most of the remaining chestnut growers: the "bread tree" was considered lost and the chestnut era finished. Experts recommended coppicing the orchard stands (the young sprouts were initially considered as more resistant to the disease; Biraghi 1953) or underplanting them with rapidly growing conifers. Paradoxically, the possibility of selling the felled old and declining trees to tannin factories provided an additional source of income and thus a further stimulus for leaving the mountain regions. This abandonment mainly affected the upper Apennines, where most of the chestnut orchards were devoted to flour production. On the contrary, the orchards producing *marroni* continued to be cultivated, at least to some extent.

15.5. The turning point of the 1980s

In the 1980s, after decades of almost total abandonment of chestnut culture culminating in the 1960s and 1970s, the situation began to change. The appearance and natural spread of the chestnut blight hypovirulence, first observed in Italy at the beginning of the 1950s (Biraghi 1953), allowed the chestnut orchard trees to survive and even to recover from the blight attacks. This fungal infection, defined as a healing canker, was able to colonize the bark and girdle the stem or the branch but remained on the superficial layer of the bark, and the flow of the sap was not affected, leaving the tree untouched and living. Over time, the healing canker evolved and completely healed the original infection (Turchetti et al. 2008).

Contrary to what happened with the American chestnut (Griffin 2000), the natural diffusion of the hypovirulence and the healing form of the chestnut blight gave rise to a new and very optimistic scenario for the European chestnut and its cultivation, particularly in Italy.

Moreover, some emigrants returned home after retirement and began restoring the chestnut orchards and introducing new practical and technological management and marketing strategies, including the birth of growers' associations to sustain the sector. The restoring approach consisted of removing the invading vegetation and pruning the old and grafted chestnut trees. New plantations and, more especially, conversions of coppices into orchards by means of grafting the resprouts (Gherardi et al. 1991) also took place, resulting in a significant increase in chestnut cultivation. In this context, the commercial and valuable *marroni* played a central driving role thanks to their appreciation in urban markets as a high-quality natural product and the related optimal price/performance ratio. Furthermore, when selling chestnut fruits, people also market the traditional culture and the related bucolic landscape, which is indeed far removed from the original and grim reality of the chestnut as the tree of bread. As a result, the restoring activities mostly involved the *marroni* orchards in the lowlands of the Apennines. Interestingly, this happened much less in the upper part of the mountains, where the traditional staple food orchards are located.

15.6. Current constraints and future perspectives

The chestnut tree is currently facing old and new threats (Figure 15.3). Fortunately, the hypovirulence of the chestnut blight is still predominant throughout the whole chestnut range, permitting the survival of trees and their cultivation (Pezzi et al. 2011, 2020a). The ink disease is still widespread in the originally affected areas (unpublished data), but a further spread in terms of new foci is very limited and localized (Ambrosini et al. 1997; Pezzi et al. 2011). This suggests that the pathogen does not have the potential to spread into the whole chestnut area as previously hypothesized. A new threat to the chestnut tree appeared in Europe at the beginning of the 21st century (Brusino et al. 2002) and spread to the Bolognese Apennines starting from 2008 (Graziosi and Santi, 2008): the Asian chestnut gall wasp (*Dryocosmus kuriphilus* Yasumatsu). Even if no trees were directly killed by this pest, the heavy damages to the foliage caused a drastic reduction in fruit production, again turning the confidence of the growers into very pessimistic scenarios. The biological control options with the specific parasitoid *Torymus sinensis* Kamijo (Quacchia et al. 2008) have fortunately proved to be very effective since the biological agent was first released in 2010 (Vai et al. 2014). Damage caused by the chestnut gall wasp is at present very limited and does not affect the trees' vitality or the fruit production (Pezzi et al. 2020a). Therefore, the management of the main chestnut phytosanitary issues can currently be carried out in a completely biological and sustainable way.

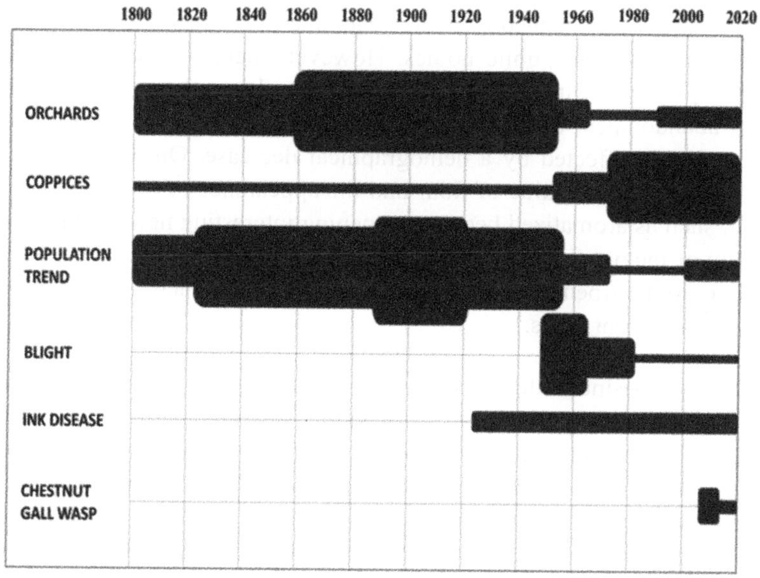

Figure 15.3. *Summary of the historical evolution of chestnut orchards and its relationship with the different constraints in the Bolognese Apennines (modified from Pezzi et al. 2011)*

A further new risk factor for the chestnut is climate change in general and the ensuing climatic extremes such as summer drought in particular (Conedera et al. 2010). Drought affecting the fruit production occurred in this part of the Apennines in 2017 and 2019. In addition, the chestnut groves were affected by a snowstorm in 2017. Such repeated climatic stress can reduce the resistance of the trees against diseases or cause localized decline phenomena. In this context, the nut rot fungus (*Gnomoniopsis castaneae* Tamietti) newly appeared, severely affecting the fruit production (especially in 2018). Its spread seems to be strictly related to the temperature increase during the vegetative season and particularly at the time of the fruit harvest.

From a socio-economic point of view, the restoration and maintenance of chestnut orchards and their ecological value is closely linked to the presence of interested owners carrying out traditional management. At present, 115 growers mainly located in the lower parts of the mountains and dedicated to *marroni* valorization are reported as affiliated to the local chestnut growers' association. Interestingly, a generational change is taking place, with an increasing number of young members. In the upper part of the range, there is a small consortium of about 30 members dedicated to chestnut cultivation and flour production. The number of consortia members has changed over the years due to the possibility of easy access to financial support from public bodies. However, there is also the effect of the aging of the growers, with an old generation not always being replaced and the subsequent abandonment of the orchards (Pezzi et al. 2017). In addition, the upper Apennines are still affected by a demographical decrease. On the other hand, the recent rediscovery of the value of flour and the appearance of new products related to chestnut, such as aromatized beers, are opening interesting new scenarios also for the economy of the old flour orchards. Nevertheless, the firms related to agriculture or silviculture in the Apennines show a constant trend of reduction, although slightly slowing down in recent years.

In summary, chestnut cultivation is still vital, but at risk of marginalization and subjected to a balancing act between abandonment and making the recognized multifunctionality potential for the mountains effective. Although the traditional management of chestnut orchards allows a high level of biodiversity to be preserved, there is still a substantial risk of abandonment of these activities, which is related to the low marketing possibilities of products, the highly fragmented property systems, and the aging of the growers and difficulties in granting a management succession. The recognition of the indispensable contribution afforded by chestnut growers should lead to the definition of political choices for territorial governance aimed at supporting, also financially, the work of these subjects. Currently, there is still an important and significant economic income in some areas, but often carried out through personal passion or transmitted between generations as a family cultural heritage. Conserving the multifunctional values of chestnut groves in the Bolognese Apennines thus calls for a complex management strategy that comprises: (i) supporting young growers and helping them to gain possession of chestnut orchards; (ii) supporting associations that enhance the political weight of growers and contribute to spreading traditional and innovative knowledge on chestnut orchards management; (iii) increasing research on both the physiology and ecology of the chestnut and the orchard trees in particular, in order to better cope with future challenges such as climate change and evolving socio-economic conditions, and (iv) exploring new strategies to give added value to the chestnut fruit products in the context of a sustainable management.

A specific effort to safeguard old-growth chestnut orchards could provide a representative example and encourage increased attention to chestnut cultivation, assuring the revitalization of this fundamental heritage of the Italian mountains.

15.7. References

Alessandri, S., Krznar, M., Ajolfi, D., Cabrer, A.M.R., Pereira-Lorenzo, S., Dondini, L. (2020). Genetic diversity of *Castanea sativa* Mill. accessions from the Tuscan-Emilian Apennines and Emilia Romagna Region (Italy). *Agronomy*, 10, 1–11.

Ambrosini, I., Gherardi, L., Viti, M.L., Maresi, G., Turchetti, T. (1997). Monitoring diseases of chestnut stands by small format aerial photography. *Geocarto International*, 12(3), 41–46.

Biraghi A. (1946). Il cancro del castagno causato da *Endothia parasitica*. *L'Italia Agric.*, 7, 406.

Biraghi, A. (1953). Possible active resistance to *Endothia parasitica* in *Castanea sativa*. *Proceedings. of the XI Congress of the International Union of Forest Research Organizations*, 643–645.

Bounous, G. (2014). *Il castagno. Risorsa multifunzionale in Italia e nel mondo*. Edagricole, Bologna.

Brussino, G., Bosio, G., Baudini, M., Giordano, R., Ramello, F., Melika, G. (2002). Pericoloso insetto esotico per il castagno europeo. *L'Informatore Agrario*, 37, 59–61.

Conedera, M. and Krebs, P. (2008). History, present situation and perspective of chestnut cultivation in Europe. In *Proceedings of the Second Iberian Congress on Chestnut*, Abreu, C.G, Peixoto, F.P., Gomes-Laranjo J. (eds). ISHS, Section Nuts and Mediterranean Climate Fruits, Leuven.

Conedera, M., Krebs, P., Tinner, W., Pradella, M., Torriani, D. (2004). The cultivation of *Castanea sativa* (Mill.) in Europe, from its origin to its diffusion on a continental scale. *Veg. Hist. Archaeobot.*, 13, 161–179.

Conedera, M., Barthold, F., Torriani, D., Pezzatti, G.B. (2010). Drought sensitivity of *Castanea sativa*: Case study of summer 2003 in the Southern Alps. *Acta Hortic.*, 866, 297–302.

Ferretti, F., Sboarina, C., Tattoni, C., Vitti, A., Zatelli, P., Geri, F., Pompei, E., Ciolli, M. (2018). The 1936 Italian kingdom forest map reviewed: A dataset for landscape and ecological research. *Ann. Silvic. Res.*, 42(1), 3–19.

Gabrielli, A. (1994). La civiltà del castagno. *Monti e boschi*, 65, 3.

Gallingani M.A. (2019). Rapporto Appennino 2019 [Online]. Available at: http://inumeridibo lognametropolitana.it/studi-e-ricerche/rapporto-appennino-2019) [Accessed 27 January 2021]

Gherardi, L., Maetzke, F., Maresi, G. (1991). Prove sperimentali di recupero produttivo dei castagneti da frutto nella collina bolognese. *L'Italia Forestale e Montana*, 46(2), 159–173.

Graziosi, I. and Santi, F. (2008). Chestnut gall wasp (*Dryocosmus kuriphilus*): Spreading in Italy and new records in Bologna province. *Bull. Insectol.*, 61(2), 343–348.

Griffin, G.J. (2000). Blight control and restoration of the American chestnut. *J. Forest.*, 98(2), 22–27.

Krebs, P., Moretti, M., Conedera, M. (2008). Castagni monumentali nella Svizzera sudalpina. Importanza geostorica, valore ecologico e condizioni sanitarie. *Sherwood*, 14(1), 5–10.

Mariotti, B., Castellotti, T., Conedera, M., Corona, P., Manetti, M.C., Romano, R., Tani, A., Maltoni, A. (2019). *Linee guida per la gestione selvicolturale dei castagneti da frutto.* Rete Rurale Nazionale 2014–2020, Scheda 22.2 – Foreste, CREA, Rome.

Morelli, F., Python, A., Pezzatti, G.B., Moretti, M. (2019). Bird response to woody pastoral management of ancient chestnut orchards: A case study from the southern Alps. *For. Ecol. Manage.*, 453, 117560.

Obrist, M.K., Rathey, E., Bontadina, F., Martinoli, A., Conedera, M., Christe, P., Moretti, M. (2011). Response of bat species to sylvo-pastoral abandonment. *Forest Ecol. Manag.*, 261(3), 789–798.

Petri, L. (1917). *Studi sulla malattia del castagno detta "dell'inchiostro".* Ricci, M., Florence.

Pezzi, G., Maresi, G., Conedera, M., Ferrari, C. (2011). Woody species composition of chestnut stands in the Northern Apennines: The result of 200 years of changes in land use. *Landsc. Ecol.*, 26(10), 1463–1476.

Pezzi, G., Lucchi, E., Maresi, G., Ferretti, F., Viaggi, D., Frascaroli, F. (2017). Abandonment or survival? Understanding the future of *Castanea sativa* stands in function of local attitude (Northern Apennine, Italy). *Land Use Policy*, 61, 564–574.

Pezzi, G., Gambini, S., Buldrini, F., Ferretti, F., Muzzi, E., Maresi, G., Nascimbene, J. (2020a). Contrasting patterns of tree features, lichen, and plant diversity in managed and abandoned old-growth chestnut orchards of the northern Apennines (Italy). *Forest Ecol. Manage.*, 470–471, 118207.

Pezzi, G., Donati, D., Muzzi, E., Conedera, M., Krebs, P. (2020b). Using chorographic sources to reconstruct past agro-forestry systems. A methodological approach based on the study case of the northern Apennines. *Landscape Res.*, 45, 359–376.

Quacchia, A., Moriya S., Bosio, G., Scapin, I., Alma, A. (2008). Rearing, release and settlement prospect in Italy of *Torymus sinensis*, the biological control agent of the chestnut gall wasp *Dryocosmus kuriphilus*. *Biol. Control*, 53, 829–839.

Quattrocchi, G. (1938). *Il miglioramento dei castagneti dell'Appennino Bolognese.* Stabilimento grafico F. Lega, Faenza.

Squatriti, P. (2013). *Landscape and Change in Early Medieval Italy: Chestnuts, Economy, and Culture.* Cambridge University Press, Cambridge and New York.

Turchetti, T., Ferretti, F., Maresi, G. (2008). Natural spread of *Cryphonectria parasitica* and persistence of hypovirulence in three Italian coppiced chestnut stands. *Forest Pathol.*, 38, 227–243.

Vai, N., Colla, R., Mazzoli, L., Bariselli, M. (2014). The regional project for biological control of the Chinese gall wasp *Dryocosmus kuriphilus* in Emilia-Romagna. In *Conference and Abstracts Book of the European Conference of Arboriculture – Planning the Green City: Relationships Between Trees and Infrastructures*, Giordano, L., Ferrini, F., Gonthier, P. (eds). DISAFA Editions, Grugliasco.

Vigiani, D. (1908). *Il castagno*. Casa Agricola Fratelli Ottavi, Casale Monferrato.

Zagnoni R. (1997). La coltivazione del castagno nella montagna fra Bologna e Pistoia nei secoli XI–XIII. Atti delle Giornate di Studio (Capugnano, 14 settembre 1996), Gruppo di studi alta valle del Reno-Società pistoiese di storia patria ("Storia e ricerca sul campo fra Emilia e Toscana", 5), 41–57.

16

Claudius' Coin in the Forest – Niche Construction and Strategies by Early Colonizers of Boreal Inlands in Central Scandinavia

Ove ERIKSSON[1] and Karl-Johan LINDHOLM[2]

[1] *Stockholm University, Sweden*

[2] *Uppsala University, Sweden*

16.1. Introduction

At the time when the first emperors ruled the Roman Empire around 2,000 years ago, a colonization process commenced in the vast boreal forest inland regions of central Scandinavia (Figure 16.1). The people involved in this colonization may have lived in the remote outskirts of the inhabited world, but they had knowledge of agriculture and iron production technology, and they most likely held a world-view not so different from other people closer to the Roman frontier. Perhaps they had heard something about a mighty empire far away in the south; at least someone had, as a coin from the time of Emperor Claudius (AD 41–54) has been retrieved from a lump of earth found in a slope towards the river Indalsälven near the small village of Krångede (Figure 16.1). In this chapter, we examine this colonization, focusing on the challenges faced by people, the strategies used to meet these challenges, and how the colonization altered the landscape and vegetation. Due to the still limited evidence, our examination should be regarded as tentative and our conclusion as a

Historical Ecology,

hypothesis. The time frame is from early Iron Age (the Scandinavian Iron Age is usually defined as the period 500 BC–AD 1050) until the early Middle Ages, but we also briefly remark on later history and legacies in the present-day landscape in the region.

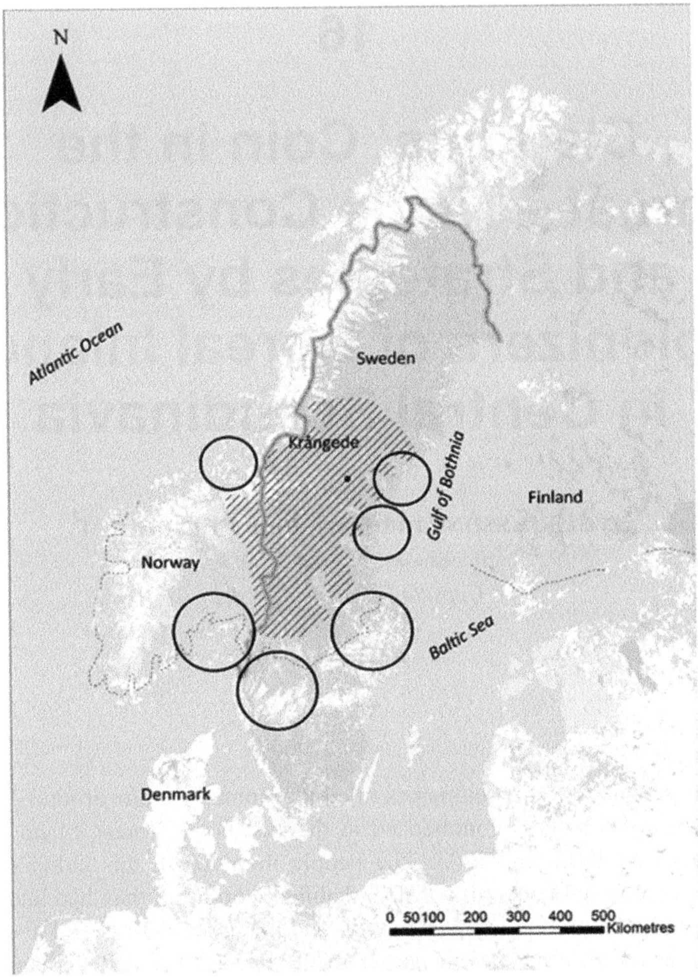

Figure 16.1. *A map showing Scandinavia, with the "boreal forest inlands regions of central Scandinavia" hatched. The circles mark the closest main agricultural regions during the early Iron Age. The dotted line marks the southern border of present-day boreal zone. The dot marks Krångede, the place where Claudius' coin was found. For a color version of this figure, see www.iste.co.uk/decocq/ecology.zip*

In southern and south-eastern Scandinavia about 2,000 years ago, there were extensive cultural landscapes with roots in the Neolithic, several thousand years earlier. Forests had been cleared, and particularly on the plains, landscapes were more or less fully exploited, with settlements, fields, meadows and pastures. Iron tools were commonly available. At this time, land-use was organized according to so-called "infield systems". The infields – i.e. settlements, crop fields and meadows – were protected (by fences or stone walls) from uncontrolled grazing by livestock, and surrounded by "outland", mainly used for livestock grazing and collection of various resources (Pedersen and Widgren 2011; Eriksson and Arnell 2017; Eriksson et al. 2021). During the first centuries AD, several "power centers" developed, probably local "elite" chiefdoms, from which the colonization processes emanated, for example south-eastern Sweden, the Gulf of Bothnia and western Norway (Figure 16.1).

Figure 16.2. *A typical view of the boreal forest inland region of central Scandinavia, showing a shieling (secondary farm) in the province of Jämtland. Photo by Andreas Hennius. For a color version of this figure, see www.iste.co.uk/decocq/ecology.zip*

In conventional historical narratives, the boreal forest inland regions of central Scandinavia have been considered as being more or less unaffected by people until the Middle Ages (e.g. Myrdal 2011), only inhabited by sparse populations of hunters. This view has been questioned, and there is evidence that the regions were colonized much earlier (e.g. Magnusson 1986; Lindholm et al. 2013) and became integrated parts of trade networks (e.g. von Stedingk and Baudou 2006), primarily with iron (Hyenstrand 1979; Magnusson 1986; af Geijerstam and Nisser 2011), and also with products from hunting (Lindholm and Ljungkvist 2016; Hennius 2020) and tar (Hennius 2018). Similarly, in biology, a common narrative is that boreal inland forests were largely "untouched" until what we may call modern forestry began during the 19th century (see Eriksson (2018), for a review). Several in-depth studies of forest history have shown that this narrative is flawed (e.g. Josefsson et al. 2010). However, the people colonizing these boreal forest inland regions during the Iron Age and the Middle Ages would have faced great challenges. The climate is harsh and except for some areas, for example along rivers and plains surrounding large lakes, most land consists of coniferous forest and mires not primarily suitable for agriculture (Figure 16.2). The main objective of this chapter is to examine the strategies people used to handle these challenges, and how these affected the landscape.

16.2. Concepts and theoretical framework

In our studies of the colonization of the boreal forest inland, we use a conceptual and theoretical framework, human niche construction, useful for examining continuous interactions between humans and their environment. The underlying idea is that human actions, intentional or unintentional, alter the environment in a way that feedback to people, either enabling them to increase their resource base, promote inventions or constraining their further actions. Over time, these continuous interactions may lead to cultural "adaptations", in this context meaning perceptions and practices which promote the persistence of local societies (for an extensive discussion of these concepts, see Eriksson et al. 2018). Niche construction can be defined broadly as the processes by which a species (e.g. humans) alters its own ecological niche, or the niche of other species, and the feedback of these alterations to the niche constructing species (Odling-Smee et al. 2003). The "ecological niche" here means the totality of the resources and environmental requirements used by the niche constructing species. "Niche" therefore equals "the environment", or some part of it. A key point in niche construction is that the species not only respond to the environment (as it is often considered), but also in fact creates its own environment, for good, or for bad.

Niche construction theory was originally developed in biology (Odling-Smee et al. 2003), but has increasingly been applied to interactions between humans and environment (e.g. Laland and O'Brien 2012; Eriksson et al. 2021), for example concerning domestication of plants and animals (e.g. Zeder 2016). Indeed, on a large spatial scale, we may consider human-mediated transformation of the environment as a domestication of whole landscapes. Over time there is an "ecological memory" of past environmental alterations, and this ecological memory is what studies of current landscapes and vegetation recognize as legacies of previous land-use.

16.3. A historical overview of the colonization

It is likely that the initial driver behind the colonization of the boreal forest inland regions during the first centuries AD was iron production. Iron was produced in bloomery furnaces using lake or bog ore (af Geijerstam and Nisser 2011). In just one of the present-day provinces of the region (Jämtland-Härjedalen), over 700 iron production sites have been found, and it is likely that the true number is higher (Magnusson 1986). Based on remains of slag, Magnusson (1986) found that the earliest iron production sites in the boreal forest inland regions (up until around the mid first millennium AD) were considerably larger than later production sites (from AD 900 onwards). This suggests that there was an "overproduction" of iron used for trade during the first centuries AD, in line with what would be expected if the colonization was initiated, directly or indirectly, by the Iron Age "chiefdoms" located west, east and southeast of the inland forest regions (Figure 16.1). After a period of decline around the 7th–8th century AD, the colonization of the boreal forest inland regions increased during late Iron (Viking) Age and early Middle Ages (e.g. Myrdal 2011), probably at least partly driven by the same mechanism, iron production used for trade. However, the development of the blast furnace technology during the early Middle Ages made other regions more efficient in producing iron, and the smaller iron production sites in the inland forest regions at that time indicate that they may have served only local demands. The colonization was halted during the "late medieval agrarian crisis", which followed after a series of plague epidemics (mainly the "Black Death" in mid-14th century), wars and climate cooling. Many farms were abandoned. Along with the recovery from these societal crises, the inland forests in central Sweden may have been particularly attractive to colonizers, since previously abandoned farms in more central regions had been used for pasture or meadow, and thus became unavailable to people when the population started to increase (Myrdal 2011). During the 18th and 19th centuries, the number of small farms increased as a result of a growing population and initiatives by the Swedish State to promote agricultural expansion (Gadd 2011). Many of these small farms were later abandoned, as rural people moved into urban centers for work in industries.

After this brief overview, we return to the earlier phases of the colonization process, initiated during the Iron Age. Before focusing on the strategies employed by people, it is relevant to consider what kind of knowledge, and what kind of world-view, "cosmology", the colonizers may have brought with them. It is likely that the people colonizing the boreal forest inland regions of Scandinavia had similar knowledge and world-view as people from the main agricultural regions (Lindholm et al. 2013). It seems evident that people must have known iron production technology as the earliest iron production sites were located further to the south in comparison with the forest inland regions (af Geijerstam and Nisser 2011). Furthermore, infield systems developed the centuries before and after AD (Eriksson 2020). These systems included the conception of land as infields (fields, meadows), probably "owned" by persons or families, and outland, which was used as commons, shared by several farms. Furthermore, livestock (cattle, sheep and goats) were the basis of the agricultural system, since manure was needed to fertilize the small permanent crop fields. Livestock was commonly kept indoors ("stalled"), parts of the year. Stalling may have been introduced for various reasons, for example protection from theft and to enhance collection of manure, but may eventually have served to protect livestock from harsh winters (Eriksson 2020). Thus, fodder production on meadows (hay or leaf-hay) was an essential component of the agricultural system. From more recent times, we know the yearly cycle of management of wooded meadows adapted to maximize the production of hay (including leaf-hay), but it is likely that these management practices were established during the early Iron Age, along with the invention of scythes and hay-rakes useful for hay harvest (Eriksson 2020). The links from plant biomass (fodder), not accessible as food for humans, via manure, to fertilization of crop fields, were understood (although "understood" would mean something very different for people 2,000 years ago, who had no knowledge about chemistry or nutrients).

16.4. A structured landscape

16.4.1. *Constructing the environment*

Several components of infield systems were literally the result of niche construction, for example crop production on manured permanent fields and meadow management, and these formed a basis of conceptions of how land was managed and organized, and probably also about who had the rights to use the land, an early form of "private ownership" (Eriksson and Arnell 2017). Although hunting, for example Eurasian elk and reindeer (Hennius 2020), and fishing were probably part of the basis for subsistence during the colonization, people came from agricultural regions, i.e. they were farmers. We should imagine that people involved in the colonization of the inland forests had a mind-set based on infield systems with their constructed production sites, fields and meadows, and where stalling of livestock was part of the yearly cycle, and where herded livestock during the vegetation period grazed in semi-open forests (as they become when livestock grazes forests).

These people now faced a landscape with large tracts of boreal forest and mires and where only minor areas along rivers and large lakes were initially suitable for agriculture. A constraint was also that iron production sites had to be chosen where iron ore is available, and these sites were not necessarily most suitable for settlements (Magnusson 1986). Thus, the use of the landscape had to be spatially structured into areas for subsistence based on agriculture and areas for iron production. We suggest that the structuring of the land-use, which initially was a "problem" to solve, represents an overarching strategy employed by people to sustain their livelihood under these conditions. The landscape as a whole became domesticated, and different components of this domestication were result of niche construction. We stress that when we use the term "strategy", we do not imply that this was planned with foresight, that anyone cleverly thought it out from the beginning. Strategies develop over time, as a result of a combination of an initial mind-set, necessities due to constraints, accumulated experience and knowledge transfer, and they may have been partly unintentional byproducts of decisions initially made for other purposes.

16.4.2. *Managing livestock*

Forest grazing by livestock has been used as long as agriculture has existed in Scandinavia, and it was a component of infield systems as well, i.e. grazing was conducted in the outland, mostly forests. Unmanaged boreal forests are not very suitable for livestock grazing, as the major tree species, Scots pine (*Pinus sylvestris*) and Norway spruce (*Picea abies*), and the dominating field layer composed of ericaceous shrubs are not sufficient feed for cattle and sheep; goats may have thrived better, and it is possible that goats were mostly used initially. However, if the forest is cleared, for example for collection of wood for fuel or charcoal, and especially if it is burned, more attractive feed is promoted, such as the grass *Deschampsia flexuosa*, herbs such as *Chamaenerion angustifolium*, and deciduous species such as birch (*Betula* spp.), goat willow (*Salix caprea*) and rowan (*Sorbus aucuparia*) (e.g. Groven and Niklasson 2005). Burning to clear forests and improve forest grazing is likely to have been used since long in southern Sweden. This management may therefore have been part of the colonizer's initial knowledge, although it was here placed in a different environmental context. The natural fire frequency of boreal forests would thus have increased due to anthropogenic impact (e.g. Granström and Niklasson 2008).

For southern Sweden during the 17th century onwards, Dahlström (2006) estimated that livestock density in forests could be as low as c. 0.1 cattle/hectare. The figure is likely to be even lower in northern boreal forest because these are less productive. Considering that cattle which graze in forests move up to eight kilometers per day, large areas had to be burnt in order to sustain even quite small herds of livestock. For climatic reasons in the forest inland regions, livestock needed to be hosted indoors during the coldest and wettest parts of the year, probably at least seven to eight months. Large quantities of fodder were needed. Initially, and for large parts of Scandinavia, leaf-hay was used as fodder, i.e. leaves and twigs from deciduous tree species (Slotte 2001). As the dominating tree species in boreal forest, Scots pine and Norway spruce, are poor as fodder, also fodder production required active manipulation of the forest environment. Again, burning would be the best way to achieve a decent production of fodder, both hay (grass, herbs) and leaf-hay. However, this required that the area used for fodder production could not simultaneously be used for grazing. As stalling of livestock was done at the farm, and in order to reduce the need of hay-transport, it is reasonable that the area closest to the settlement was allocated to fodder production. This means that the constructed landscape resembled the structure of infield systems, although the spatial scale was considerably larger, as the overall productivity of the boreal forest is so much smaller than on the agricultural plains. Livestock should then be herded at a distance, perhaps even a considerable distance, from the settlements.

In the boreal forest landscape, mires (including bogs) cover vast areas, and these provide an additional source of fodder, mainly sedges (*Carex* spp.). Management of outland hay-meadows on mires is well known from more recent times, when management often included manipulation of the water-regime in order to promote hay production (e.g. Elveland 2015). Harvesting hay on wetlands is extremely labor-intensive, and the manpower available for such harvest is likely to have been a constraint. Furthermore, as with iron production sites, the sites with the most useful mires may not be close to the settlements, implying a need for long transports of the hay after harvest. Thus, we recognize that people may have experienced several constraints. Herding of livestock (at distances away from settlements) and hay-harvest were time-consuming. In addition, hay harvested far from settlements must either be transported directly after harvest or stored until transport during winter. These activities necessitated an extension of the spatial scale of the area needed for the capturing of resources needed for subsistence; here, we also remind that iron production sites were located away from settlements. We suggest that these constraints led to the development of an overarching strategy for subsistence in the boreal forest inland: shielings.

16.4.3. *Shielings (secondary farms)*

Shielings are secondary farms, belonging to a main farm, but located at some distance away and used during parts of the year, i.e. a kind of transhumance system where some people and most livestock moved away from the main farm in order to extend the resource base for subsistence (e.g. Svensson 2018). Shielings are usually placed in the context of farming systems from more recent times, and based on a 19th century model of shielings, Larsson (2009) argued that these originated no earlier than the late Middle Ages. Other authors suggest that shielings originated during the Iron Age (e.g. von Stedingk and Baudou 2006). Based on the constraints described above, we agree with the latter opinion, and suggest that the places that became shielings developed as a cultural adaptation to the specific conditions facing people during colonization of the forest inland.

Equipped with a mind-set based on infield systems, people may have strived to construct the landscape in that way. However, the poor production of boreal forests necessitated clearing and burning of large areas. The area closest to settlements had to be allocated for production of leaf-hay (or hay from grass and sedges if settlements were located close to rivers). In addition, extensive areas were needed for outland grazing. It is thus easy to imagine that an idea developed to organize summer grazing at some distance from the main farm, thus relieving the herders from moving livestock daily between the main farm and the grazing grounds, and ensuring that livestock did not destroy the production of hay in the vicinity of settlements. A general explanation for the origin of shielings would be that they were adaptive in a resource-poor landscape, as a means to increase the exploitation of fodder by seasonally moving livestock beyond the areas close to the infields. This would seem as a natural thing to do, if we assume that the colonizers brought with them the knowledge of infield systems. Karlsson et al. (2010) suggested that the shieling-farm connection also served to define an area that demonstrated the extent of the farm's territory, thus preventing other colonizers from establishing there.

We still lack of clear material evidence for early shielings, but as pointed out by Lindholm et al. (2013), remains of early shielings may be archaeologically obscure. Magnusson (1986) made the interesting remark that iron production was often located at sites which seemed as suitable sites for shielings. We may speculate that the clearing of forest for producing charcoal used in iron production created preconditions for the localization of an early form of shielings; the forest areas cleared for charcoal production were used for livestock grazing. This would also have "rationalized" the guarding of the livestock herd, since people were engaged in other activities at the same site. In fact, Emanuelsson and Segerström (2002) suggested that a mutual relationship ultimately developed between charcoal production, slash-and-burn agriculture (which probably was introduced during late Middle Ages) and livestock grazing.

16.5. Concluding remarks

Figure 16.3. *A shieling from the province of Värmland. This shieling is still managed, as a biological cultural heritage. For a color version of this figure, see www.iste.co.uk/decocq/ecology.zip*

The colonization and establishment of agriculture and iron production in the boreal forest inland regions of central Scandinavia from the Iron Age onwards implied that people must have overcome great difficulties. Climate is harsh, and vast areas are covered with forest and mires initially unsuitable for farming. We suggest that a combination of a mind-set based on infield systems occurring in the regions from where the colonizers came, and constraints imposed by the environment, promoted human niche construction processes specific for these regions. The landscape was basically structured as infields and outland, but the constraints imposed by the poor productivity of the forests necessitated extending forest grazing spatially, promoted use of outland mires for harvest of winter fodder, and ultimately led to development of secondary farms, shielings, as an innovative strategy of a spatially structured domesticated landscape. Although initially developed during the Iron Age, other processes transformed shielings to what they became later during history. Overlaid with more recent use of boreal forests, mainly modern forestry, there are still remaining legacies in the landscape, for example remaining shielings (Figure 16.3), and rare cases of a forest structure that mirrors former use of forests as grazing ground, with large Scots pines, grass and herbaceous plants (e.g. Ericsson et al. 2000). The concept of the boreal forest regions as a cultural landscape has not been much acknowledged in research, and we end with a hope that this landscape's more than 2,000-year long history of human impacts should receive the attention it deserves.

16.6. References

Dahlström, A. (2006). Pastures, livestock number and grazing pressure 1620–1850. Ecological aspects of grazing history in south-central Sweden. Thesis, Acta Universitatis Agriculturaea Suecica 2006:95 [In Swedish with English summary].

Elveland, J. (2015). Wet hay-meadows in northern Sweden – Their history and present status. *Svensk Bot. Tidskr.*, 109, 292–336 [In Swedish with English summary].

Emanuelsson, M. and Segerström, U. (2002). Medieval slash-and-burn cultivation: Strategic or adapted land use in the Swedish mining district. *Environ. Hist.*, 8, 173–196.

Ericsson, S., Östlund, L., Axelsson, A.-L. (2000). A forest of grazing and logging: Deforestation and reforestation history of a boreal landscape in central Sweden. *New Forests*, 19, 227–240.

Eriksson, O. (2018). What is biological cultural heritage and why should we care about it? An example from Swedish rural landscapes and forests. *Nat. Conserv.*, 28, 1–32.

Eriksson, O. (2020). Origin and development of managed meadows in Sweden: A review. *Rural Landsc. – Soc., Environ., Hist.*, 7(2), 1–23.

Eriksson, O. and Arnell, M. (2017). Niche construction, entanglement and landscape domestication in Scandinavian infield systems. *Landsc. Res.*, 42, 78–88.

Eriksson, O., Ekblom, A., Lane, P., Lennartsson, T., Lindholm, K.-J. (2018). Concepts for integrated research in historical ecology. In *Issues and Concepts in Historical Ecology: The Past and Future of Landscapes and Regions*, Crumley, C.L., Lennartsson, T., Westin, A. (eds). Cambridge University Press, Cambridge.

Eriksson, O., Arnell, M., Lindholm, K.-J. (2021). Historical ecology of Scandinavian infield systems. *Sustainability*, 13, 817.

Gadd, C.-J. (2011). The agricultural revolution in Sweden 1700–1870. In *The Agrarian History of Sweden: 4000 BC to AD 2000*, Myrdal, J. and Morell, M. (eds). Nordic Academic Press, Lund.

af Geijerstam, J. and Nisser, M. (eds) (2011). *Swedish Mining and Metalworking*. National Atlas of Sweden. Norstedts Förlagsgrupp, Stockholm.

Granström, A. and Niklasson, M. (2008). Potentials and limitations for human control over historic fire regimes in the boreal forest. *Phil. Trans. R. Soc. B*, 363, 2353–2358.

Groven, R. and Niklasson, M. (2005). Anthropogenic impact of past and present fire regimes in a boreal forest landscape of southeastern Norway. *Can. J. Forest Res.*, 35, 2719–2726.

Hennius, A. (2018). Viking Age tar production and outland exploitation. *Antiquity*, 92, 1349–1361.

Hennius, A. (2020). Towards a refined chronology of prehistoric pitfall hunting in Sweden. *Eur. J. Archaeol.*, 23, 530–546.

Hyenstrand, Å. (1979). Iron and iron economy in Sweden. In *Iron and Man in Prehistoric Sweden*, Clarke, H. (ed.). Jernkontoret, LTs Förlag, Stockholm.

Josefsson, T., Gunnarsson, B., Liedgren, L., Bergman, I., Östlund, L. (2010). Historical human influence on forest composition and structure in boreal Fennoscandia. *Can. J. Forest Res.*, 40, 872–884.

Karlsson, H., Emanuelsson, M., Segerström, U. (2010). The history of a farm-shieling system in the central Swedish forest region. *Veg. Hist. Archaeobot.*, 19, 103–119.

Laland, K.N. and O'Brien, M.J. (2012). Cultural niche construction: An introduction. *Biol. Theory*, 6, 191–202.

Larsson, J. (2009). Summer farms in Sweden 1550 to 1920: An important element in North Sweden's agricultural system. Thesis, Jamtli Förlag, Östersund [In Swedish with English summary].

Lindholm, K.-J. and Ljungkvist, J. (2016). The bear in the grave: Exploitation of top predator and herbivore resources in first millennium Sweden – First trends from a long-term research project. *Eur. J. Archaeol.*, 19, 3–27.

Lindholm, K.-J., Sandström, E., Ekman, A.-K. (2013). The archaeology of the commons. *J. Archaeol. Anc. Hist.*, 10, 1–49.

Magnusson, G. (1986). *Prehistoric and Medieval Iron Industry in the Province of Jämtland*. Jernkontorets Bergshistoriska Skriftserie N:r 22. Jernkontoret, Stockholm [In Swedish with English summary].

Myrdal, J. (2011). Farming and feudalism 1000–1700. In *The Agrarian History of Sweden: 4000 BC to AD 2000*, Myrdal, J. and Morell, M. (eds). Nordic Academic Press, Lund.

Odling-Smee, F.J., Laland, K.N., Feldman, M.W. (2003). *Niche Construction: The Neglected Process in Evolution*. Princeton University Press, Princeton.

Pedersen, E.A. and Widgren, M. (2011). Agriculture in Sweden, 800 BC – AD 1000. In *The Agrarian History of Sweden: 4000 BC to AD 2000*, Myrdal, J. and Morell, M. (eds). Nordic Academic Press, Lund.

Slotte, H. (2001). Harvesting of leaf-hay shaped the Swedish landscape. *Landsc. Ecol.*, 16, 691–702.

von Stedingk, H. and Baudou, E. (2006). Capitalism in central Norrland, Sweden, during the Iron Age. *Curr. Sw. Archaeol.*, 14, 177–198.

Svensson, E. (2018). The scandinavian shieling – Between innovation and tradition. In *Historical Archaeologies of Transhumance across Europe*, Costello, E. and Svensson, E. (eds). Routledge, London.

Zeder, M.A. (2016). Domestication as a model system for niche construction theory. *Evol. Ecol.*, 30, 325–348.

17

Recent History of Vegetation Changes in the Arctic

Antoine BECKER-SCARPITTA[1,2], Bastien PARISY[1] and Tomas ROSLIN[1,3]

[1] *University of Helsinki, Finland*

[2] *Institute of Botany of the Czech Academy of Science, Průhonice, Czech Republic*

[3] *Swedish University of Agricultural Sciences, Uppsala, Sweden*

17.1. Introduction

While human-induced global change is affecting all regions on Earth, it is in the Arctic that its effects on ecological communities and native populations are currently best visible (Post et al. 2019). There, the climate is warming at a rate of nearly twice the global average, with a direct effect on permafrost, both sea and land surface ice, the snow dynamic, and the functioning of the ecosystem (Box et al. 2019). Understanding and describing what is happening in the Arctic can thus help inform future projections for the rest of the Earth. The Arctic region is also home to several indigenous human populations, who – being, for example reindeer herders or hunter-gatherers – depend on the local vegetation. A sudden change in vegetation has direct consequences for their lifestyles, migration patterns and survival.

To reveal the effects of climate on ecosystems, the archives of historical ecology – i.e. observations made over the past decades – provide an unparalleled resource. In this chapter, we start with a brief description of the Arctic tundra biome, and then summarize how historical data have been used to describe and understand recent trends in Arctic tundra vegetation. In doing so, we focus on changes observed during the last five decades, while refraining from digging into the paleoecological records.

Historical Ecology,
coordinated by Guillaume DECOCQ.

17.2. The Arctic tundra biome

Stretching north from the latitudinal tree line, the Arctic region is characterized by tundra vegetation – i.e. low-productivity rate vegetation mainly composed of herbaceous plants, shrubs, bryophytes and lichens. In total, the Arctic tundra biome spans nearly 7 million km^2, and is characterized by harsh environmental conditions. Winters are long and cold, and the growing season short, relatively cold, and typically dry. Direct human activity in the area is generally low, with the exception of herbivory pressure from reindeer herding.

Climate warming in the Arctic is proceeding at a rapid pace. While global surface temperatures have increased by 2°C on average in recent decades, annual mean temperatures in the Arctic region have risen by an average of 4°C – whereas wintertime temperatures have increased by up to 7°C (Box et al. 2019). The faster and greater change in winter, as compared to other seasons, has resulted in earlier snowmelt, lower levels of ice and snow cover and major biological changes (Arctic Report Card 2020 – https://www.arctic.noaa.gov/). At the same time, the Arctic region has experienced a general increase in air humidity and precipitation – although with high local variations. Moreover, under all climatic scenarios, the Arctic will continue warming more rapidly than any other region on Earth (IPCC 2014).

The current pattern of warming is being imposed on a flora adapted to harsh climatic conditions, with typical plant traits including short stature, waxy leaves and dense leaf hairs. Tundra plants typically form a dense mat of vegetation, or compact, cushion-like clusters, and their roots tend to grow along the soil surface, thereby taking advantage of the warmest and best-drained layers of the soil. Species diversity is low, and much of the biome is dominated by a relatively few species, each of which is widely distributed across the biome (Walker et al. 2018). We can distinguish five major growth forms among tundra plants, based on their physical, ecological and biogeochemical characteristics. Described progressing from warmer to colder conditions, they are: (i) low- and high-shrub, (ii) erect dwarf-shrubs, (iii) prostrate dwarf-shrubs, (iv) forb-lichen-moss cushions and (v) graminoids and forbs, which grow under the driest soil conditions (Raynolds et al. 2019).

17.3. The Arctic historical ecological archive

The historical ecological records amassed since the 20th century provide valuable data for distinguishing the drivers of population and community responses to global change (Vellend et al. 2013). This historical data is derived from old satellite images or photographs, the long-term monitoring of permanent plots and semi-permanent vegetation plots, and repeated visits to previously surveyed sites. Given the relatively short history of arctic ecology, we will mainly focus on records from the 1960s and later.

17.3.1. *Remote sensing over time*

The remoteness of the Arctic biome imposes both logistical and financial challenges for anyone wanting to conduct field research, especially across multiple sites. Thus, remote sensing methods are of paramount importance for studying arctic change at the landscape or regional scale (Beamish et al. 2020). Of these methods, satellite-based imagery provides some of the most powerful testimonies of ecological changes (Myers-Smith et al. 2020). Another important remote sensing method is the use of historical aerial photographs. Such imagery provides information that can complement satellite data, and allows us to see back in time to the pre-satellite era. Satellite imagery can be used to quantify different types of vegetation through their spectral reflectance signatures. The dominant metric used over the last four decades is the Normalized Difference Vegetation Index (NDVI), which is based on the reflectance signature of the vegetation. Since being developed in the late 1980s, NDVI has been widely used as a proxy for measurement of photosynthetic activity and vegetation productivity in the Arctic.

The use of NDVI is not without complications, since the spectral signature of plants may be comingled with other abiotic factors at the landscape scale, and since the interpretation of images is subject to influence from the methodological choices made. Satellite images are sensitive to clouds and snow cover, as well as to water retention and topography. From a methodological perspective, the resolution and accuracy of the images may vary with the sensor type and satellite technology used, with the algorithms applied to the images, and with the specific timing of the observations. Overall, the very nature of the arctic biome offers several challenges to studying it, since radiation enters at a low angle in the region, the growing season is short, and landscapes are typically heterogeneous (Beamish et al. 2020).

17.3.2. *Field-based records*

17.3.2.1. *Long-term monitoring of permanent plots*

To work across multiple remote sites, arctic ecologists early on joined forces in an effort called the International Tundra Experiment (ITEX; https://www.gvsu.edu/itex/). Established in the 1990s, this initiative is based on a standardized vegetation monitoring protocol that is designed to be implemented by different researchers at different sites. A primary objective of ITEX has been to manipulate temperature with open-top chambers (OTCs), to thereby test the effect of warming on tundra plant phenology, growth, community composition and diversity. To distinguish the specific effect of the warming treatments, the protocol also includes the monitoring of control plots.

Spanning over 30 years of standardized plot-level studies, the ITEX data now provide some of the largest, most comprehensive and longest time-series of vegetation change in the arctic biome. Other local long-term monitoring initiatives span different areas and time periods, such as the Greenland Ecological Monitoring (GEM – https://data.g-e-m.dk/) program. In the nature reserves of the former USSR, communities and seasonal events have been systematically recorded for decades, and even sometimes for over a century (Ovaskainen et al. 2020).

17.3.2.2. *Resurveys of semi-permanent vegetation plots*

In the 1960s, the main objective of plant ecology was typically to classify vegetation in relation to local environmental descriptors. Such sampling campaigns were mainly aimed at a type of vegetation inventory called phytosociology (or geobotany). A valuable legacy of these campaigns is the standardized protocol that they developed. As their plots can oftentimes be relocated and resampled decades later, ecologists can then infer what changes have occurred in the intervening period. Although legacy studies in the Arctic have been limited to a relatively small number of sites, they are of great value in testing hypotheses about the environmental drivers of vegetation-level change (Callaghan et al. 2011).

17.3.2.3. *Historic field observations and herbaria*

The ecological records held in museums, in herbaria, and at research stations can offer valuable sources of information on historical community composition, phenology and population dynamics. While observations and experiments from the Arctic tend to be clumped together in space and to be brief in duration, herbarium specimens collected over centuries allow us to explore another dimension of long-term ecological change (Lang et al. 2019). Over the last two decades, museums across the world have initiated the digitization of their specimens. Now the resulting open access data allows ecologists to resolve ecological questions. In the Arctic, the combination of different datasets from herbarium specimens, photographs and field observations have allowed novel descriptions of the changes taking place in tundra plant phenology in response to warming (Panchen and Gorelick 2017). However, unlike semi-permanent plots, which can be relocated with some confidence, botanical descriptions of an area are rarely associated with exact locations. This weakens the ecological conclusions that can be derived from this type of data.

17.4. Changes over time in tundra vegetation

In this section, we provide an overview of the recent changes in arctic tundra vegetation revealed by the historical archives detailed above. During the last few decades, we have seen three major dimensions of ecological change, namely, changes in (1) vegetation productivity, (2) vegetation phenology and (3) plant community structure, composition and diversity.

17.4.1. *Changes in vegetation productivity*

Large-scale analyses of satellite images indicate a general, long-term increase in NDVI in the Arctic tundra region. This trend has been termed "Arctic greening", referring to an increase in plant productivity. Conversely, a decrease in NDVI is referred to as "browning", suggesting a decrease in plant productivity or vegetation cover. The pattern of greening was first described over the period 1981–1991, and is now considered one of the clearest continental-scale responses to recent human-induced global warming. The underlying mechanisms of greening appear to be: (i) an increase in existing shrub patches, commonly refer to as expansion, (ii) an increase in plant productivity, including an increase in plant height and leaf area and (iii) a shift in species distribution, mainly due to colonization (Myers-Smith et al. 2011).

Although a growing body of evidence has linked human-induced warming over the past 40 years to the decadal change in vegetation productivity, the direction and intensity of this change have been far from uniform across the Arctic (Huang et al. 2017; Berner et al. 2020). Over the past two decades, several sources indicate a reversal in the direction of NDVI change in the Arctic biome, from greening to browning. Indeed, since the 2000s the global increase in NDVI has slowed. In north-eastern Eurasia, and northern and northwestern America, it appears to have even decreased. Several active research programs are currently exploring regional differences in patterns of productivity change, attempting to explain the causes. In particular, increased snow cover in the winter and a late spring snowmelt have been suggested as potential explanations (Bhatt et al. 2013). On top of these directional trends in NDVI, topographic position, the composition of the vegetation (e.g. shrub cover), water balance and snowmelt dynamics appear to generate additional geographic variation in plant productivity (Reichle et al. 2018).

17.4.2. *Changes in vegetation phenology*

Phenology refers to the timing of recurrent life cycle events in living organisms. Plant phenology includes, for example, the timing of flowering, seed maturation, bud break or senescence. For most plant species, phenological events are often closely linked to temperature or photoperiod. Thus, in the context of global climate change, changes in phenology are a key indicator of a species response to warming (Thackeray et al. 2016). In recent decades, remarkable changes have been observed in the phenology of plants in the Arctic tundra biome. These observations derive from long-term plot-level monitoring and experiments (Prevéy et al. 2019), from remote sensing (Beamish et al. 2020), and from herbarium specimens, historical photographs and field observations (Panchen and Gorelick 2017).

17.4.2.1. *Changes in timing*

Beyond providing a measure of primary productivity, data on NDVI provide a useful measure of the time of the season when this productivity occurs, and thus of phenology. Time series of satellite-derived NDVI have been widely used to detect an earlier start and a later end of the growing season, and thus of a lengthening growing season (Zeng et al 2013). Warming experiments conducted at the ITEX sites suggest an advancement in the timing of both leaf emergence and flowering time, whereas they show no change in or a slight delay of leaf senescence (Bjorkman et al. 2020). However, the direction and magnitude of phenological changes vary considerably between biomes, and within a biome between different species (Roslin et al. 2021).

17.4.2.2. *Shortening flowering seasons*

In an Arctic-wide synthesis, Prevéy et al. (2019) used records from long-term monitoring and warming experiments to test for a relationship between the temperature sensitivity and phenological niches of 253 tundra species. What these authors detected was a greater change in late-flowering species than in early-flowering species. Thus, they concluded that the community-wide flowering season is contracting in response to climate warming, and that this contraction is due to an advance in the flowering date of late-flowering species – without corresponding changes in early-flowering species. Such a shortening of the flowering season is consistent with other local studies in the Arctic region (Høye et al. 2013). Although increasing air temperature has a direct effect on local phenology, the snow cover modulates the effect of air temperature on conditions in the soil. Thus, snow dynamics modify the relationship between temperatures and realized phenology (Kelsey et al. 2021).

17.4.2.3. *Cascading effects on trophic interactions*

A shorter flowering season will directly limit resource availability for higher trophic levels (e.g. pollinators or herbivores), thereby affecting food web structure (Hegland et al. 2009). Since the growing season in the Arctic region is short to begin with, different trophic levels are generally well synchronized in time. Contrasting shifts in phenology between plants and higher trophic levels might then lead to a phenological mismatch (Høye et al. 2013). In a resource-limited environment such as the Arctic tundra, we may also see competition for pollinators among plants. Climate change can thereby result in a negative feedback loop from pollinators back to the plant community (Schmidt et al. 2017).

17.4.3. *Changes in plant community structure, composition and diversity*

Environmental changes in the Arctic have resulted in shifts in community composition, in species' abundance and distribution, and in plant functional traits (Nabe-Nielsen et al. 2017; Bjorkman et al. 2018a). As remote sensing methods have yet to reach the accuracy needed to reliably detect such small-scale changes, relevant historical data are therefore mainly limited to field observations. In this context, the ITEX site network provides the most extensive and longest-term monitoring of community composition in the tundra biome. A synthesis using data from 46 individual ITEX control plots (1980–2010) revealed a relationship between temperature warming and changes in community composition, despite high variability between sites (Elmendorf et al. 2012). What was specifically found was an increase in the size of herbaceous plants, in the organic layer of the soil, and in the abundance of shrubs and herbaceous plants. At the same time, the area of bare soil was reduced.

Complementing the findings from the ITEX sites, the "Back to the Future" research program has generated repeated observations of selected local communities. Similar patterns emerged across the individual sites that were revisited: a general sensitivity of tundra vegetation to warming and to changing precipitation regimes, an increase in shrub abundance and a shift in shrub distribution. Equally important is the finding that patterns differ considerably between sites, highlighting a strong context-dependence in the responses of tundra plant communities to warming (Callaghan et al. 2011).

17.4.3.1. *Change in functional traits*

Until recently, the exploration of functional traits among Arctic plants was mostly restricted to coarse functional grouping based on growth forms such as shrubs, forbs, graminoids, mosses and lichens. Such crude classification of plants by their appearances is hardly linkable to key ecosystem functions. Yet, in 2018, the publication of the database generated by the Tundra Trait Team (TTT) marked a milestone in the history of Arctic trait-based ecology. This database builds on field measurement of 18 key functional traits for 978 tundra species, which traits can be directly linked to key ecosystem functions and assembly rules (Bjorkman et al. 2018b). This conquering of the functional dimension allows ecologists to test hypotheses related to how current change is affecting community and ecosystem functioning at a biogeographical scale.

By using the plant Tundra Trait database, Bjorkman et al. (2018a) quantified changes in intraspecific variability and community-level variability over time. Consistent with several studies showing increases in plant productivity and growth rate (see above), the trait-based approach reveals an increase in weighted community height across the biome over the past three decades, while other traits (e.g. leaf traits) appear not to have changed over time. The observed change in plant community height seems partly related to changes in abiotic conditions and to changes in species composition, rather than to changes in species abundance. In other words, the observed change in vegetation height reflects the range expansion of taller species towards higher latitudes and higher elevations in response to warming.

17.4.3.2. *Shrubification*

A major change in the functional composition or growth form of tundra plants is the expansion of shrubs. Using historical oblique aerial photographic tracking from 1945, Tape et al. (2006) demonstrated that shrubs have expanded over a large area in Alaska. Another example of the use of historical aerial photography is the study by Frost and Epstein (2014), who took advantage of the declassification of high-resolution photographs of the former USSR, taken in the 1960s by US intelligence, to demonstrate that shrubs and trees have expanded in the tundra ecotone since that time.

As mentioned above, a general greening of the tundra (see section 17.4.1) has been attributed to shrub expansion. In terms of community structure, shrubification is mainly attributed to a positive response to warming and to changing precipitation regimes by genera such as *Betula*, *Salix* and *Alnus*, and also by *Empetrum*, *Cassiope* and others. The trends observed by remote sensing are supported by in situ observations of changes in shrub abundance, of shifts in community composition and in species distribution, and of increases in annual growth rings and in plant size (Myers-Smith et al. 2019). The expansion of shrubs across the tundra has direct consequences for ecosystem functioning, as it may inter alia decrease albedo, thus resulting in altered snow accumulation and melting dynamics, changes in carbon cycling by increasing photosynthetic activity, and changes in the thickness of the organic layer. However, shrubification may also have direct consequences for biotic communities, causing changes in understory species composition through changes in canopy conditions (i.e. in light, temperature and humidity), and changes in the herbivory regime and/or in microbial and fungal activity (Myers-Smith et al. 2011).

17.5. Synthesis and perspectives

A growing body of evidence links human-induced warming over the past four decades to changes in the species composition, productivity and structure of arctic vegetation. However, the direction and trends of this change vary across space, time and taxa (Huang et al. 2017). Using three major ecological patterns of change – productivity, phenology and community composition – we have demonstrated the value of different historical ecological archives for identifying and understanding these changes. Our synthesis highlights that imprints are most clearly detected when complementary data of multiple types are considered. Yet, importantly, such comparisons also expose conflicts in the evidence provided by different types of historical data. As an example of consistent trends exposed by multiple types of evidence, the greening trend observed over the tundra biome – and more recently, the browning observed within given regions – was suggested by remote sensing but later verified through in situ observations of shrubification in the Arctic landscape. A large body of evidence now supports an increase in plant height and shrub abundance, in the face of limited temporal change in vegetation composition. As an example of conflicting evidence, remote sensing methods have pointed to an earlier start and later end of the growing season, resulting in a longer growing season. By comparison, long-term in situ monitoring of phenology has revealed a general shortening of the flowering season.

Importantly, the major trends of vegetation change detected across the Arctic are almost invariably associated with a high variability in the direction and magnitude of change across space, over time and among species. This points to a strong context-dependence in how tundra vegetation responds to climate change. The spatial heterogeneity of ecological changes in the tundra vegetation underlines the need for widespread observations conducted under variable conditions. One major challenge then is the highly uneven spatial coverage of the current ecological archive. There is thus an urgent need to extend current research to new sites, thereby improving our understanding of arctic complexity.

As discussed elsewhere in this book, and specifically for the arctic region in this chapter, historical records provide a crucial avenue towards detecting long-term ecological change. Every arctic study, and in fact every observation, has the potential to contribute to the historical archives of the decades to come. It is therefore the responsibility of every contemporary scientist to structure their data in standard formats and to store them in openly accessible databases – thereby benefitting their future peers and colleagues.

17.6. References

Arctic Report Card (2020). Arctic Program [Online]. Available at: https://www.arctic.noaa.gov/ [Accessed March 2021].

Beamish, A. et al. (2020). Recent trends and remaining challenges for optical remote sensing of Arctic tundra vegetation: A review and outlook. *Remote Sensing of Environment*, 246, 111872 [Online]. Available at: https://doi.org/10.1016/j.rse.2020.111872.

Berner, L.T. et al. (2020). Summer warming explains widespread but not uniform greening in the Arctic tundra biome. *Nature Communications*, 11(1), 4621. doi: 10.1038/s41467-020-18479-5.

Bhatt, U. et al. (2013). Recent declines in warming and vegetation greening trends over pan-arctic tundra. *Remote Sensing*, 5(9), 4229–4254. doi: 10.3390/rs5094229.

Bjorkman, A.D. et al. (2018a). Plant functional trait change across a warming tundra biome. *Nature*, 562(7725), 57–62. doi: 10.1038/s41586-018-0563-7.

Bjorkman, A.D. et al. (2018b). Tundra trait team: A database of plant traits spanning the tundra biome. *Global Ecology and Biogeography*, 27(12), 1402–1411. doi: 10.1111/geb.12821.

Bjorkman, A.D. et al. (2020). Status and trends in arctic vegetation: Evidence from experimental warming and long-term monitoring. *Ambio*, 49(3), 678–692. doi: 10.1007/s13280-019-01161-6.

Box, J.E. et al. (2019). Key indicators of arctic climate change: 1971–2017. *Environmental Research Letters*, 14(4), 045010. doi: 10.1088/1748-9326/aafc1b.

Callaghan, T.V. et al. (2011). Multi-decadal changes in tundra environments and ecosystems: Synthesis of the international polar year-back to the future project (IPY-BTF). *Ambio*, 40(6), 705–716. doi: 10.1007/s13280-011-0179-8.

Elmendorf, S.C. et al. (2012). Plot-scale evidence of tundra vegetation change and links to recent summer warming. *Nature Climate Change*, 2(6), 453–457. doi: 10.1038/nclimate1465.

Frost, G.V. and Epstein, H.E. (2014). Tall shrub and tree expansion in Siberian tundra ecotones since the 1960s. *Global Change Biology*, 20(4), 1264–1277. doi: 10.1111/gcb.12406.

GEM (2021). The Greenland Ecosystem Monitoring (GEM) Database [Online]. Available at: https://data.g-e-m.dk/ [Accessed March 2021].

Hegland, S.J. et al. (2009). How does climate warming affect plant-pollinator interactions? *Ecology Letters*, 12(2), 184–195. doi: 10.1111/j.1461-0248.2008.01269.x.

Høye, T.T. et al. (2013). Shorter flowering seasons and declining abundance of flower visitors in a warmer Arctic. *Nature Climate Change*, 3(8), 759–763. doi: 10.1038/nclimate1909.

Huang, M. et al. (2017). Velocity of change in vegetation productivity over northern high latitudes. *Nature Ecology & Evolution*, 1(11), 1649–1654. doi: 10.1038/s41559-017-0328-y.

IPCC (2014). Climate Change 2014: Synthesis Report. Contribution of Working Groups I, II and III to the Fifth Assessment Report of the Intergovernmental Panel on Climate Change. Geneva [Online]. Available at: https://www.ipcc.ch/report/ar5/syr/.

ITEX (2021). International Tundra Experiment (ITEX) [Online]. Available at: https://www.gvsu.edu/itex/ [Accessed March 2021].

Kelsey, K.C. et al. (2021). Winter snow and spring temperature have differential effects on vegetation phenology and productivity across Arctic plant communities. *Global Change Biology*, 27(8), 1572–1586. doi: 10.1111/gcb.15505.

Lang, P.L.M. et al. (2019). Using herbaria to study global environmental change. *New Phytologist*, 221(1), 110–122. doi: 10.1111/nph.15401.

Myers-Smith, I.H. et al. (2011). Shrub expansion in tundra ecosystems: Dynamics, impacts and research priorities. *Environmental Research Letters*, 6(4), 045509. doi: 10.1088/1748-9326/6/4/045509.

Myers-Smith, I.H. et al. (2019). Eighteen years of ecological monitoring reveals multiple lines of evidence for tundra vegetation change. *Ecological Monographs*, 89(2), e01351. doi: 10.1002/ecm.1351.

Myers-Smith, I.H. et al. (2020). Complexity revealed in the greening of the Arctic. *Nature Climate Change*, 10(2), 106–117. doi: 10.1038/s41558-019-0688-1.

Nabe-Nielsen, J. et al. (2017). Plant community composition and species richness in the High Arctic tundra: From the present to the future. *Ecology and Evolution*, 7(23), 10233–10242. doi: 10.1002/ece3.3496.

Ovaskainen, O. et al. (2020). Chronicles of nature calendar, a long-term and large-scale multitaxon database on phenology. *Scientific Data*, 7(1), 47. doi: 10.1038/s41597-020-0376-z.

Panchen, Z.A. and Gorelick, R. (2017). Prediction of Arctic plant phenological sensitivity to climate change from historical records. *Ecology and Evolution*, 7(5), 1325–1338. doi: 10.1002/ece3.2702.

Post, E. et al. (2019). The polar regions in a 2°C warmer world. *Science Advances*, 5(12), 1–12. doi: 10.1126/sciadv.aaw9883.

Prevéy, J.S. et al. (2019). Warming shortens flowering seasons of tundra plant communities. *Nature Ecology & Evolution*, 3(1), 45–52. doi: 10.1038/s41559-018-0745-6.

Raynolds, M.K. et al. (2019). A raster version of the circumpolar arctic vegetation map (CAVM). *Remote Sensing of Environment*, 232(Oct), 111297. doi: 10.1016/j.rse.2019.111297.

Reichle, L.M. et al. (2018). Spatial heterogeneity of the temporal dynamics of arctic tundra vegetation. *Geophysical Research Letters*, 45(17), 9206–9215. doi: 10.1029/2018GL078820.

Roslin, T. et al. (2021). Phenological shifts of abiotic events, producers and consumers across a continent. *Nature Climate Change*, 11(3), 241–248. doi: 10.1038/s41558-020-00967-7.

Schmidt, N.M. et al. (2017). Interaction webs in arctic ecosystems: Determinants of arctic change? *Ambio*, 46(S1), 12–25. doi: 10.1007/s13280-016-0862-x.

Tape, K. et al. (2006). The evidence for shrub expansion in Northern Alaska and the Pan-Arctic. *Global Change Biology*, 12(4), 686–702. doi: 10.1111/j.1365-2486.2006.01128.x.

Thackeray, S.J. et al. (2016). Phenological sensitivity to climate across taxa and trophic levels. *Nature*, 535(7611), 241–245. doi: 10.1038/nature18608.

Vellend, M. et al. (2013). Historical ecology: Using unconventional data sources to test for effects of global environmental change. *American Journal of Botany*, 100(7), 1294–1305. doi: 10.3732/ajb.1200503.

Walker, D.A. et al. (2018). Circumpolar arctic vegetation classification. *Phytocoenologia*, 48(2), 181–201. doi: 10.1127/phyto/2017/0192.

Zeng, H. et al. (2013). Shifts in arctic phenology in response to climate and anthropogenic factors as detected from multiple satellite time series. *Environmental Research Letters*, 8(3), 035036. doi: 10.1088/1748-9326/8/3/035036.

18

Reconstructing the Impact of Humans on Aotearoa New Zealand's Biodiversity

Nicolas J. RAWLENCE[1], Alexander J.F. VERRY[1,2], Karen GREIG[1], Justin J. MAXWELL[3], Lara D. SHEPHERD[4] and Richard WALTER[1]

[1] *University of Otago, Dunedin, New Zealand*

[2] *University of Toulouse, France*

[3] *Sunrise Archaeology, Mangonui, New Zealand*

[4] *Museum of New Zealand Te Papa Tongarewa, Wellington, New Zealand*

18.1. Introduction

Aotearoa New Zealand's (NZ) biodiversity has been described as "… *close as we will get to the opportunity to study life on another planet*" (Diamond 1990). As an isolated archipelago, its biodiversity has been shaped by both tens of millions of years of isolation and frequent long-distance dispersal. A dynamic geological history – including partial Oligocene inundation, tectonic uplift, Pleistocene glacial-interglacial cycles and volcanism – has also played a large role in the evolution of NZ's biota. At the time of Polynesian arrival, NZ's vertebrate fauna was dominated by birds and reptiles, which exhibited a range of island adaptations including gigantism, flightlessness and K-selected breeding strategies (Worthy and Holdaway 2002). There were at least 217 avian species breeding in the NZ ecological region at the time of human arrival, ranging in size from giant moa (Dinornithiformes) and the world's largest eagle (Haast's eagle, *Aquila moorei*) to among the smallest of birds, the Acanthissittid wrens (Worthy et al. 2017). The herpetofauna was dominated by Eugongylinae skinks, Diplodactylid geckos, tuatara (the last surviving member of

Historical Ecology,
coordinated by Guillaume DECOCQ.

the Rhynchocephalia family) and Leiopelmatid frogs (Worthy and Holdaway 2002). NZ also had a diverse marine mammal assemblage which included multiple species of cetaceans and pinnipeds (Seersholm et al. 2018).

Paleoecological and archaeological evidence suggests Polynesians colonized NZ in the late 13th century CE (Wilmshurst et al. 2008) at a time of relative climatic stability (Rawlence et al. 2016a), followed by European settlement from 1796 CE. In addition to hunting and environmental modification (Perry et al. 2014a, 2014b), humans also introduced exotic predators (e.g. rats, dogs, mustelids), plants and changed the genetic landscape through translocations and plant cultivation (Shepherd et al. 2018). By the early 20th century, the NZ ecosystem had fundamentally changed with the extinction of ~50% of its biodiversity, biological turnover events, range contractions, population bottlenecks and significant changes in indigenous forest cover (Waters et al. 2017; Rawlence et al. 2019).

Our understanding of these dramatic changes has taken a step-change with multidisciplinary research incorporating morphological analysis of the paleontological and archaeological record, in combination with radiocarbon-dating, stable dietary isotopes and ancient DNA. This chapter focuses on the anthropogenic impacts on NZ biodiversity up until the early 20th century, from an archaeological, paleoecological and evolutionary perspective.

18.2. Archaeological evidence for anthropogenic impact in New Zealand

NZ was settled at the last phase of a maritime expansion across the Pacific. As settlers moved east across the Pacific Ocean, islands became increasingly small and isolated. There was a natural biodiversity decline to the east, although most islands supported abundant bird populations (Steadman 2006). These species evolved with few terrestrial predators and the finely balanced ecosystems were disrupted when humans arrived with domestic pigs, fowl, dogs, rats and plants, and cleared land for agriculture. Naïve avifauna, particularly flightless and ground nesting birds, were susceptible to hunting. What followed was a wave of extirpation and extinction recorded in the archaeological records of many islands (Steadman 2006).

This pattern of human impact following the establishment of horticulture and hunting, which led to rapid faunal extinctions, was repeated across Polynesia and eventually NZ. The first settlers came to NZ from East Polynesia ~1300 CE (Wilmshurst et al. 2008) from "Hawaiki", which was probably located in a zone encompassing the Cook, Austral and Society Islands. Archaeological evidence throughout the "Hawaiki Zone" shows that islands were linked into an interaction sphere; a zone of regular inter-island voyaging and exchange (Walter et al. 2017). Colonization and the subsequent pattern of networked mobility were enabled by long-distance voyaging craft, navigation skills and a portable economy – the ability to transform natural landscapes into cultural ones. NZ was settled out of this dynamic zone of interaction.

Arriving in NZ, the first settlers adopted an effective strategy for systematic exploration of coastlines, river systems and resource bases. Genetic evidence suggests a large founder population was required to account for current levels of diversity, implying multiple settlement voyages over a short period (Walter et al. 2017). By the mid-1300s CE, NZ had been explored; all major resources had been discovered and were being widely traded in coastal exchange networks (Walter et al. 2017). Settlement spread rapidly and a pattern emerged involving dispersed coastal occupation, with clusters of sites around harbors, and larger villages at river mouths.

The first settlers' strategy was to reproduce settlement patterns, economics and lifestyles of their tropical homeland; characterized by high levels of mobility facilitated by ocean-going canoes, long-distance exchange of high-quality resources, the maintenance of social relationships, and the significance of harbors, river mouths and estuaries for settlement. As in "Hawaiki", populations were dispersed (but connected) and like their Pacific ancestors, settlers took up hunting to supplement their horticultural and fishing economy for a generation or two.

Moa were the most important prey species and their significance in the Early Māori (pre-1500 CE) economy is well reflected in the archaeological record. Numerous large "moa-hunter" village sites have been excavated around river mouths in NZ, containing the remains of hundreds of moa that were hunted in the hinterlands and brought to the villages for butchering and consumption. The rich material culture and ornate burial practices of these early communities reflect a high level of wealth and security. In addition to large villages, small seasonal hunting sites and specialized processing sites have been identified, containing the butchered and cooked remains of one to two moa, plus associated stone tools (Anderson 1989). Discussion of moa exploitation revolves around the interpretation of bone assemblages from early sites. There is abundant evidence, however, that hunters were also engaged in high levels of egg predation. Most sites with moa bone also contain eggshell, often in large quantities, and a few early sites have produced whole moa eggs (Anderson 1989). As in Polynesia, NZ avifauna evolved in the absence of mammalian predators including humans. Of the 131 known endemic species present in NZ at the time of Polynesian settlement, at least 40 became extinct prior to European arrival at a time of relative climatic stability (Duncan and Blackburn 2004) as a result of different birding strategies adopted by early Māori.

There is little compelling evidence that non-moa bird hunting in NZ was ever intensive, or involved specialized, taxon-specific hunting strategies. It appears to have been largely opportunistic (Anderson 1989) and taxon-specific impacts as a result of hunting can be reasonably predicted on the basis of habitat, flight ability, body size and nesting behavior (Duncan and Blackburn 2004). The archaeological record documents the scale, range and impact of hunting, but little is understood about hunting practices, strategies and the actual process by which targeted faunas were affected. Some models of hunting behaviors leading to moa extinction have been proposed, including Anderson's (1989) discussion of rolling wave, blitzkrieg or serial overkill scenarios. But these, and other models of overkill, assume a lengthy period of hunting, with colonization up to 1,000 years ago. With refinements in radiocarbon-dating it is now apparent that settlement of NZ occurred ~1300 CE (Wilmshurst et al. 2008). Furthermore, Perry et al. (2014a) demonstrated that moa were extinct no more than 150 years later, and perhaps less. The problem of moa extinction rates has been addressed in genetic models (e.g. Allentoft et al. 2014), but there is almost nothing concrete known about moa hunting practices from the archaeological record. So far, no hunting tools have been identified and there is no evidence on bones (e.g. impact trauma or dog tooth marks) that might indicate a killing method. Whatever answer, in terms of per-annum kill rate, that ecological modeling might predict for these K-selected birds, this is only a small part of the puzzle. The bigger question revolves around behavioral aspects of predation – what were Māori actually doing to have such a rapid impact on moa. The situation is no less enigmatic for the problem of pinniped hunting.

NZ's coastline had abundant fur seal (*Arctocephalus forsteri*) and sea lion (*Phocarctos hookeri*) colonies on both main islands (Smith 1989). By European contact, fur seals were restricted to the southwestern South Island, and sea lions and southern elephant seals (*Mirounga leonina*) were extirpated from mainland NZ (Rawlence et al. 2016a). Pinnipeds have been reported from early Māori sites around NZ but are absent from North Island sites by 1500 CE and decline sharply after that in the south with cryptic biological turnover events (Waters et al. 2017). Smith (1989) reconstructed two strategies adopted by Māori hunters: regular seasonal cropping of breeding colonies and opportunistic killing of beached individuals. The former was the most economically important and had the highest impact, the latter was less frequent but occurred more widely throughout NZ.

Unlike moa hunting, there are very few sites with abundant pinniped bone and thus little direct evidence for large-scale hunting. Instead, the evidence most frequently cited that Māori heavily impacted pinnipeds is the reduction in their range. The presence of breeding colonies in northern NZ at the time of human arrival (Smith 1989) implies a significant range contraction, as a result of early

Māori predation. In fact, it implies a greater range contraction than that caused by 18th century CE industrial sealing. So far, predation on that scale is not directly supported by archaeological evidence. However, genetic modeling indicates sea lion extirpation is possible even with low but sustained levels of hunting (e.g. < 0.5 individuals/person/year; Rawlence et al. 2016a).

18.3. Paleovegetation change in pre- and post-European contact New Zealand

Environmental modification was an important part of the successful colonization of Polynesia, and fire was the principal tool to modify vegetation communities. In NZ, it has been argued that large-scale vegetation change occurred rapidly. The range of hypothesized drivers for change include improving the landscape for cultivation, collection or encouragement of wild crops, or to remove dense forest and improve travel or resource access (McGlone et al. 2005). Vegetation modifications may have been unintentional, resulting from other activities.

Palynology and charcoal analysis have been the main methods used to study vegetation change through time. Most pollen records are used to investigate natural processes of long-term, multi-millennial paleoenvironmental change, addressing various themes especially climate change. These studies can also inform on fire regimes, both locally and nationally; changes in sedimentation regimes; and the timing and impact of human arrival, although anthropogenic changes are often only a small component of this type of research.

It has been estimated that 85% of NZ's landmass was covered in forest before humans arrived (Perry et al. 2014b). Native forests varied greatly due to differences in climate and topography across NZ. While forests were diverse, they generally shared the following characteristics during the Holocene: low fire frequency, a majority of endemic species not adapted to fire and, as a result, a majority of ecosystems highly vulnerable to fire (Perry et al. 2014b).

Polynesian arrival coincided with a major change in fire frequency and intensity. The initial burning period (IBP) was widespread and occurred within a few centuries of human arrival (Perry et al. 2014b). The combination of forest species not adapted to fire, human-lit fires and vegetation changes which occurred after firing all increased flammability of these environments. This led to long-term environmental modifications of vast areas, with loss during the IBP of 40% of the existing forests and most of the lowland and montane dry forests which were the most fire susceptible (Perry et al. 2014b).

A plant that is admirably adapted to fire is the bracken fern (*Pteridium esculentum*). Bracken recolonizes fire-affected areas, has a high reproductive rate and survives repeated firing. It is not surprising then that after human arrival, bracken fern became one of the most abundant species in pollen records and, combined with charcoal, is often used as a proxy for pre-European anthropogenic environmental modification. Bracken fern also produces large, edible rhizomes, which were an important carbohydrate in the Māori diet. Edible bracken was mainly recovered from deep, fertile soils and the presence of bracken in other locations has been hypothesized to represent environmental management to improve traveling conditions (McGlone et al. 2005).

Investigations of NZ paleovegetation, primarily designed to study pre-European contact human activity are less common. A paleobotanical study of human-generated vegetation change on Rēkohu (Chatham Island, one of NZ's offshore islands), which was settled by Moriori ~1450 CE, combined traditional anthropological and anthracological (charcoal) datasets, enhancing our understanding of human impact on small islands. Vegetation changes were rapid and long-lasting. The Rēkohu pollen and charcoal record indicated that soon after human arrival, fires resulted in long-term loss of much of the forest and a significant increase in bracken fern. However, ethnographic, archaeological and anthracological data, combined with palynology, indicated that coastal forests survived and were heavily managed in a form of arboriculture (Maxwell et al. 2016). The primary species in these managed forests was kōpi/karaka (*Corynocarpus laevigatus*), which was translocated from the NZ mainland. Karaka pollen preserves poorly and is rarely noted in pollen records, and is probably highly underrepresented (Maxwell and Tromp 2016). Palynological records alone therefore could not have provided a reliable representation of this species in the past, nor could they have contributed to our understanding of its relative abundance or its importance to Moriori subsistence.

From 1769 CE, European arrival initiated what would be a second wave of large-scale vegetation change in NZ. There was further forest loss and substantial changes occurred when new food plants were introduced. The earliest industries in post-European contact NZ centered on the harvest of natural resources from sea and land. On land, forests were harvested from the 1820s CE to supply timber spars for the British Navy, from the 1830s CE to supply cut timber to Australia and beyond and, as the population of NZ increased, to supply building timber locally (King 2003).

From the 1840s CE onward, there was also widespread loss of forest for agriculture. It is estimated that most of the forest loss (~90%) occurred early in the European period, between 1830 and 1873 CE, including the widespread loss of old-growth forests (Perry et al. 2014b). Fires that occurred at this time were large and often not intentionally set, as accounts often cite a combination of drought conditions and specific ignition sources. Today, less than 25% of the land area of NZ is forested.

18.4. Utilizing Aotearoa's natural resources: Māori cultivation and translocation of flora and fauna

As Polynesians colonized the Pacific, they took useful plants and animals with them. At least some of these were introduced to NZ during settlement, for example, kūmara (*Ipomoea batatas*) and kurī, the Polynesian dog (*Canis familiaris*). However, many of the plants, which were important food staples in the tropical Pacific, failed to establish or only flourished in warmer parts of NZ. Māori and Moriori brought into cultivation a number of indigenous plant species, and it is also suspected that they translocated several mollusks and fish far outside their natural distribution to secure a more predictable resource supply (e.g. Maxwell and Tromp 2016; Ross et al. 2018). Given its late settlement, NZ offers a unique opportunity to examine the early stages of cultivation and domestication (e.g. what stage in the domestication process is genetic diversity lost), in conjunction with Māori oral histories, which complement western scientific methods and provide a more complete picture of past translocations.

There is evidence of selection for morphological characters in plants. Some karaka trees growing near former Māori habitation sites have larger fruit compared with wild trees (Maxwell and Tromp 2016). There was selection of a number of cultivars of tī kōuka/cabbage tree (*Cordyline australis*), whose roots and stems were used as a carbohydrate source. The main tī kōuka cultivar in the North Island was tī para, a sterile form that could be easily propagated vegetatively. A number of plants were grown for their fiber; the most prized of these was harakeke/NZ flax (*Phormium tenax*). At least 60 cultivars of harakeke were developed for different fiber characteristics (Wehi and Clarkson 2007).

Māori also greatly extended the distributions of some plants and animals through translocations to new regions. Some movements were recorded in myth and oral histories (see below), but others have been inferred based on unusual species distributions including occurrence around sites of former Māori occupation. Powell (1938) suggested that several flax snail (*Placostylus* spp.) populations around past Māori occupation sites may have resulted from translocation. Karaka trees were introduced outside their natural range in northern North Island into southern North Island and northern South Island, as well as to the distant Kermadec and Chatham Islands (800 km and 650 km from the mainland, respectively). This is supported by oral histories about the origins of karaka groves (Atherton et al. 2015). There are myths relating to some of the founding ancestors, that talk about the seeding of rivers and lakes with indigenous species of fish and invertebrates, including kōaro (*Galaxias brevipinnis*), kōura (*Paranephrops planifrons*) and kākahi (*Echyridella* spp.; McDowall 2011).

Genetic data has been used to examine past Māori translocations. High levels of genetic structuring within natural populations of rengarenga (*Arthropodium cirratum*), an endemic plant with edible rhizomes, allowed inference that it had been brought into cultivation at least three times from different source populations (Shepherd et al. 2018). Similarly, genetic relationships between populations of tītīrangi (*Veronica speciosa*) indicate that the sole extant South Island population probably derived from human-mediated dispersal from one of the most distant natural populations (Armstrong and de Lange 2005). Genetic data from karaka and the surf clam toheroa (*Paphies ventricosa*) revealed low genetic diversity and a lack of structuring in natural populations resulting in the source(s) of species translocations being unclear (Atherton et al. 2015; Ross et al. 2018). High-throughput sequencing methods may provide greater power for distinguishing between such populations but have been little used for reconstructing translocation histories in NZ to date. Other challenges for reconstructing Māori translocations include distinguishing between natural and cultivated populations in species with wide natural distributions and the impact of hybridization and European translocations on genetic patterns. Techniques such as palynology, anthrocology, radiocarbon-dating and paleoenvironmental DNA sampling of ancient sediments can also inform about translocations, especially in determining the timing of introductions (Maxwell et al. 2016).

Translocation of plants and animals by Māori contributed to the modification of NZ. Such introductions led to changes in the structure of ecosystems, many of which are not yet fully understood. There is the potential for translocated species to become invasive in their new habitats (e.g. karaka). Dealing with such invasive species can be challenging because the control of their invasiveness must be balanced with preserving culturally significant plantings.

18.5. Evolutionary consequences of Polynesian and European arrival

Polynesian arrival in NZ resulted in the extinction of ~50% of animal species present at the time, including at least 54 avian species, three frogs and one skink (Worthy et al. 2017). Unique lineages of conspecific taxa also went extinct (e.g. Rawlence et al. 2016a). This was achieved through hunting, environmental modification and predation from kiore, the Pacific rat (Rattus exulans), and kurī (Greig and Rawlence 2021). One of the most well-studied extinctions is that of the moa. With stable (or increasing) population sizes when humans arrived (Allentoft et al. 2014), all nine species were extinct within 150 years (Perry et al. 2014a). Extensive radiocarbon-dating of archaeological sites also shows that several other taxa went extinct during this period. This includes Waitaha penguin (*Megadyptes waitaha*) and the prehistoric NZ, and Chatham Islands lineages of sea lion, which are only known from subfossil, and early archaeological sites (Rawlence et al. 2016a; Waters et al. 2017).

European arrival resulted in an additional wave of extinctions (~30 avian species to date including huia *Heteralocha acutirostris* and laughing owl *Ninox albifaces*) as new mammalian predators were introduced (e.g. mustelids, ship (*R. rattus*) and Norway (*R. norvegicus*) rats), alongside further environmental modification. This culminated in the arrival of ship rats on Big South Cape Island in the 1960s CE, which resulted in the extinction of Stead's bush wren (*Xenicus longipes variabilis*), South Island snipe (*Coenocorypha iredalei*) and the greater short-tailed bat (*Mystacina robusta*). One posterchild species of European impact was the relictual Lyall's wren (previously Stephen's Island wren; *Traversia lyalli*), which was extirpated from mainland NZ no doubt as a result of predation by kiore (Worthy and Holdaway 2002). Subsequent predation by feral cats (*Felis catus*) rendered the bird extinct by 1895 CE. Anthropogenic pressures facilitated by Europeans (e.g. harvesting, introduction of game fish and their diseases, habitat modification) likely led to the only known extinction of an endemic NZ freshwater fish, the NZ grayling (*Prototroctes oxyrhynchus*). The museum trade potentially also contributed to further extinctions (e.g. Auckland Island merganser *Mergus australis*), even while species protections were being put in place (Worthy and Holdaway 2002).

The faunal extinctions created vacancies in ecological niches. In some cases, ancient DNA and radiocarbon-dating has shown these niches were subsequently filled in biological turnover events by closely related taxa or a different genetic lineage of the original taxon. To date, these previously cryptic events have been observed in *Megadyptes* penguins (Waitaha to yellow-eyed *M. antipodes*), little penguins (*Eudyptula minor* to *E. novaehollandiae*), sea lions (prehistoric NZ to subantarctic lineage) and waterfowl (Poūwa *Cygnus sumnerensis* to black swan *C. atratus*; Rawlence et al. 2017; Waters et al. 2017). While the extinctions were contemporaneous with the first few centuries after human arrival, replacements in most cases were temporally offset. In contrast, *Megadayptes* penguins and sea lions exhibited synchronous biological turnover, with replacement potentially facilitated by cooler climate and reduced anthropogenic pressure during the Little Ice Age (Waters et al. 2017). Closely related, open-habitat adapted Australian avifauna also replaced forest-adapted NZ avifauna as a result of the extinction of NZ congeners and environmental modification (Rawlence et al. 2019).

While numerous NZ species went extinct, many others were reduced to relict populations, coincident in some cases with significant reductions in genetic and morphological diversity (e.g. Scarsbrook et al. 2022). One such species, the South Island takahē (*Porphyrio hochstetteri*) was considered extinct until 1948 CE when it was rediscovered. Much of NZ's herpetofauna, especially larger species (e.g. Hamilton's frog *Leiopelma hamiltoni*, tuatara *Sphenodon punctatus* and Duvaucel's gecko *Hoplodactylus duvaucelii*) found refuge on offshore islands following Polynesian arrival and mainland extirpation (Worthy and Holdaway 2002). Ancient DNA research is highlighting the impacts of these bottlenecks. kākāpō (*Strigops habroptilus*) underwent two bottlenecks associated with Polynesian, and later, European arrival (Seersholm et al. 2018). Otago shag (*Leucocarbo chalconotus*) lost 99% of its genetic variation within a century of Polynesian arrival (Rawlence et al. 2016b), while fur seals exhibited significant changes in haplotype frequency following contraction into and expansion out of refugia (Salis et al. 2016). However, not all range-contractions were genetically detrimental – Fiordland crested penguin (*Eudyptes pachyrhynchus*) retained much of its genetic diversity, despite retraction into refugia (Cole et al. 2019).

Human arrival resulted in not just extinctions, but also lost ecological connections. NZ's flora exhibits a high proportion of hypothesized anti-browsing traits (e.g. divarication, heteroblasty, spines, unpalatable leaves and toxins). Plant species with these traits have been found in moa coprolites, lending support to the anti-browsing hypothesis and loss of selective pressure post extinction (Wood et al. 2008). NZ's avifauna may also have been important pollinators and seed-dispersers. Ancient DNA and palynological analysis of kākāpō coprolites showed a high abundance of pollen from the parasitic wood rose (*Dactylanthus taylorii*), suggesting that kākāpō (now restricted to offshore islands) may have fed on and pollinated this endangered plant when compared with previously only known native pollinator, the lesser short-tailed bat (*M. tuberculata*; Wood et al. 2012). kākāpō and moa also consumed colorful fruiting ectomycorrhizal fungi and may have played a role in fungal dispersal (Boast et al. 2018). Paleodietary evidence also suggests moa were important small seed (<3.3 mm) dispersers of herbs and shrubs – large seeds do not survive passage through the gizzard (Wood et al. 2008). The extinctions and population bottlenecks experienced by NZ's avifauna may have also resulted in the co-extinction of associated parasite faunas, although further research is needed to determine if parasites found in moa and kākāpō coprolites survive in other avian herbivores (Boast et al. 2018).

18.6. Conclusion

The arrival of Polynesians, and later Europeans, in Aotearoa, set in motion a series of events that would fundamentally change the NZ ecosystem. Multidisciplinary research has been at the forefront of shedding light on this dynamic period of NZ's history, and is answering some of the most pertinent questions in archaeology and paleoecology. However, many questions remain unanswered, and await further research and new analytical techniques. As in geology, the past is the key to the present, and we would argue, also the future.

18.7. References

Allentoft, M.E., Heller, R., Oskam, C.L., Lorenzen, E.D., Hale, M.L., Gilbert, M.T.P., Jacomb, C., Holdaway, R.N., Bunce, M. (2014). Extinct New Zealand megafauna were not in decline before human colonization. *PNAS*, 111, 4922–4927.

Anderson, A. (1989). *Prodigious Birds: Moas and Moa-Hunting in Prehistoric New Zealand.* Cambridge University Press, Cambridge.

Armstrong, T.J. and de Lange, P.J. (2005). Conservation genetics of *Hebe speciosa* (Plantaginaceae) an endangered New Zealand shrub. *Bot. J. Linn. Soc.*, 149, 229–239.

Atherton, R.A., Lockhart, P.J., McLenachan, P.A., de Lange, P.J., Wagstaff, S.J., Shepherd, L.D. (2015). A molecular investigation into the origin and human-mediated spread of karaka/kōpi (*Corynocarpus laevigatus*) in New Zealand. *J. Roy. Soc. NZ.*, 45, 212–220.

Boast, A.P., Weyrich, L.S., Wood, J.R., Metcalf, J.L., Knight, R., Cooper, A. (2018). Coprolites reveal ecological interactions lost with the extinction of New Zealand birds. *PNAS*, 115, 1546–1551.

Cole, T.L., Rawlence, N.J., Dussex, N., Ellenberg, U., Houston, D.M., Mattern, T., Miskelly, C.M., Morrison, K.W., Scofield, R.P., Tennyson, A.J.D. et al. (2019). Ancient DNA of crested penguins: Testing for temporal genetic shifts in the world's most diverse penguin clade. *Mol. Phylo. Evol.*, 131, 72–79.

Diamond, J.M. (1990). New Zealand as an archipelago: An international perspective. *Ecol. Restor. NZ. Is.*, 2, 3–8.

Duncan, R.P. and Blackburn, T.M. (2004). Extinction and endemism in the New Zealand avifauna. *Global Ecol. Biogeog.*, 13, 509–517.

Greig, K. and Rawlence, N.J. (2021). The contribution of *kurī* (Polynesian dog) to the ecological impacts of the human settlement of Aotearoa New Zealand. *Front. Ecol. Evol.*, 9, 757988.

King, M. (2003). *Penguin History of New Zealand.* Penguin, Rosedale.

Maxwell, J.J. and Tromp, M. (2016). *Corynocarpus laevigatus*: Where art thou? Finding evidence of this elusive tree crop. *Rev. Palaeobot. Palynol.*, 234, 198–210.

Maxwell, J.J., Howarth, J.D., Vandergoes, M.J., Jacobsen, G.E., Barber, I.G. (2016). The timing and importance of arboriculture and agroforestry in a temperate East Polynesia society, the Moriori, Rekohu (Chatham Island). *Quat. Sci. Rev.*, 149, 306–325.

McDowall, R.M. (2011). *Ikawai: Freshwater Fishes in Māori Culture and Economy.* Canterbury University Press, Christchurch.

McGlone, M.S., Wilmshurst, J.M., Leach, H.M. (2005). An ecological and historical review of bracken (*Pteridium esculentum*) in New Zealand, and its cultural significance. *NZ. J. Ecol.*, 29, 165–184.

Perry, G.L.W., Wheeler, A.B., Wood, J.R., Wilmshurst, J.M. (2014a). A high-precision chronology for the rapid extinction of New Zealand moa (Aves, Dinornithiformes). *Quat. Sci. Rev.*, 105, 126–135.

Perry, G.L.W., Wilmshurst, J.M., McGlone, M.S. (2014b). Ecology and long-term history of fire in New Zealand. *NZ. J. Ecol.*, 38, 157–176.

Powell, A.W.B. (1938). The Paraphantidae of New Zealand, No. IV and the genus *Placostylus* in NZ. *Rec. Auck. Inst. Mus.*, 2, 141–150.

Rawlence, N.J., Collins, C.J., Anderson, C.N.K., Maxwell, J.J., Smith, I.W.G., Robertson, B.C., Knapp, M., Horsburgh, K.A., Stanton, J.A.L., Scofield, R.P. et al. (2016a). Human-mediated extirpation of the unique Chatham Islands sea lion and implications for the conservation management of remaining New Zealand sea lion populations. *Mol. Ecol.*, 25, 3950–3961.

Rawlence, N.J., Scofield, R.P., Spencer, H.G., Lalas, C., Easton, L.J., Tennyson, A.J.D., Adams, M., Pasquet, E., Fraser, C., Waters, J.M. et al. (2016b). Genetic and morphological evidence for two species of *Leucocarbo* shag (Aves, Pelecaniformes, Phalacrocoracidae) from southern South Island of New Zealand. *Zool. J. Linn. Soc.*, 177, 676–694.

Rawlence, N.J., Kardamaki, A., Easton, L.J., Tennyson, A.J.D., Scofield, R.P., Waters, J.M. (2017). Ancient DNA and morphometric analysis reveal extinction and replacement of New Zealand's unique black swans. *Proc. B.*, 284, 20170876.

Rawlence, N.J., Scofield, R.P., McGlone, M.S., Knapp, M. (2019). History repeats: Large scale synchronous biological turnover in avifauna from the Plio-Pleistocene and Late Holocene of New Zealand. *Front. Ecol. Evol.*, 7, 158.

Ross, P.M., Know, M.A., Smith, S., Smith, H., Williams, J., Hogg, I.D. (2018). Historical translocations by Māori may explain the distribution and genetic structure of a threatened surf clam in Aotearoa (New Zealand). *Sci. Rep.*, 8, 17241.

Salis, A.T., Easton, L.J., Robertson, B.C., Gemmell, N., Smith, I.W.G., Weisler, M.I., Waters, J.M., Rawlence, N.J. (2016). Myth or relict: Does ancient DNA detect the enigmatic Upland seal? *Mol. Phylo. Evol.*, 97, 101–106.

Scarsbrook, L., Verry, A.J.F., Walton, K., Hitchmough, R.A., Rawlence, N.J. (2022). Ancient mitochondrial genomes recovered from small vertebrate bones through minimally destructive DNA extraction: Phylogeography of the New Zealand gecko genus *Hoplodactylus*. *Mol. Ecol.*, doi.org/10.1111/mec.16434 [in press].

Seersholm, F.V., Cole, T.L., Grealy A., Rawlence, N.J., Greig, K., Knapp, M., Stat, M., Hansen, A.J., Easton, L.J., Shepherd, L. et al. (2018). Subsistence practices, past biodiversity, and anthropogenic impacts revealed by New Zealand-wide ancient DNA survey. *PNAS*, 115, 7771–7776.

Shepherd, L.D., Bulgarella, M., de Lange, P.J. (2018). Genetic structuring of the coastal herb *Arthropodium cirratum* (Asparagaceae) is shaped by low gene flow, hybridization and prehistoric translocation. *PLoS ONE*, 13, e0204943.

Smith, I.W.G. (1989). Maori impact on the marine megafauna: Pre-European distributions of New Zealand sea mammals. In *Saying So Doesn't Make It So: Papers in Honour of B. Foss Leach*, Sutton, D.G. (ed.). New Zealand Archeological Association, Dunedin.

Steadman, D.W. (2006). *Extinction and Biogeography of Tropical Pacific Birds*. University of Chicago Press, Chicago.

Walter, R., Buckley, H., Jacomb, C., Matisoo-Smith, E.A. (2017). Mass migration and the Polynesian settlement of New Zealand. *J. World Prehist.*, 30, 351–376.

Waters, J.M., Fraser, C.I., Maxwell, J.J., Rawlence, N.J. (2017). Did interaction between human pressure and Little Ice Age drive biological turnover in New Zealand? *J. Biogeog.*, 44, 1481–1490.

Wehi, P.M. and Clarkson, B.D. (2007). Flora of New Zealand 10. *Phormium tenax*, harakeke, New Zealand flax. *NZ. J. Bot.*, 45, 521–544.

Wilmshurst, J.M., Anderson, A.J., Higham, T.F.G., Worthy, T.H. (2008). Dating the late prehistoric dispersal of Polynesians to New Zealand using the commensal Pacific rat. *PNAS*, 105, 7676–7680.

Wood, J.R., Rawlence, N.J., Rogers, G.M., Austin, J.J., Worthy, T.H., Cooper, A. (2008). Coprolite deposits reveal the diet and ecology of the extinct New Zealand megaherbivore moa (Aves, Dinornithiformes). *Quat. Sci. Rev.*, 27, 2593–2602.

Wood, J.R., Wilmshurst, J.M., Worthy, T.H., Holzapfel, A.S., Cooper, A. (2012). A lost link between a flightless parrot and a parasitic plant and the potential role of coprolites in conservation paleobiology. *Conserv. Biol.*, 26, 1091–1099.

Worthy, T.H. and Holdaway, R.N. (2002). *The Lost World of the Moa: Prehistoric Life of New Zealand*. Canterbury University Press, Christchurch.

Worthy, T.H., De Pietri, V., Scofield, R.P. (2017). Recent advances in avian palaeobiology in New Zealand with implications for understanding New Zealand's geological, climatic and evolutionary histories. *NZ. J. Zool.*, 44, 177–211.

19

Historical Ecology of the Coastal Aeolian Sedimentary Systems of the Canary Islands

Aarón Moisés SANTANA-CORDERO, Antonio Ignacio HERNÁNDEZ-CORDERO, Néstor MARRERO-RODRÍGUEZ, Leví GARCÍA-ROMERO, Elisabet FERNÁNDEZ-CABRERA, Carolina PEÑA-ALONSO, Emma PÉREZ-CHACÓN ESPINO and Luis HERNÁNDEZ-CALVENTO

Universidad de Las Palmas de Gran Canaria, Canary Islands, Spain

19.1. Introduction

A brief history is presented in this work of the evolution of the coastal aeolian sedimentary systems of five islands of the Canary Archipelago in Spain (Tenerife, Gran Canaria, Fuerteventura, Lanzarote and La Graciosa). Taking the Spanish conquest of the islands as the starting point, the aim is to provide a synthesis of the 500-year-old relationship between Canary society and some of the island's aeolian ecosystems. This will help to explain and understand their present-day status and functioning. An analysis is undertaken of the interactions between the natural dynamics of these systems and the land-uses that have been developed in them. Historical sources were used for the period which extends from the 1500s to the 1900s, mainly archived documents, cartography and secondary sources which are

Historical Ecology,

reflected in the references. For more recent times (20th and 21st centuries), additional information was obtained from field work, historical aerial photographs and recent digital orthophotos. In both cases, the spatio-temporal analyses were principally carried out through the use of geographic information systems (GIS) to determine the environmental changes on the basis of biotic and geomorphological variables.

The role of these environments in the socioeconomic development of the Canary Islands has been crucial and it helps explaining the growing interest in recent decades in the coastal ecosystems of the Canary Islands (Spain). These systems constitute an ideal laboratory for the study of the consequences of the interactions between their natural dynamics and land-uses, particularly in reference to arid island environments. The aboriginal inhabitants of the islands had lived a pre-European existence for only around 1,500 years before they were conquered over the course of the 15th century by the Spanish. The relatively short period of human intervention since then is enriched by multiple information sources on natural features and land-uses. In short, the spatio-temporal limits are well known and the information sources varied.

19.2. Study sites

The coastal aeolian sedimentary systems occupy a total surface area of 18,810 ha (2.5% of the total land area of the archipelago) and are situated mainly on the eastern islands (Hernández-Cordero et al. 2019). The fact that these systems developed on oceanic islands, in an arid climate, and have been subjected in recent decades to intensive human activity in the form of year-round tourism is distinguishing eco-anthropic features in a European context. The climate conditions are characterized by annual precipitation of just 81–110 mm, mean annual temperatures ranging between 19.7 and 21°C, and the important influence of the trade winds which blow from the NE. The desiccating effect of the wind and the low amount of rainfall have resulted in low density and mainly shrubby vegetation (Hernández-Cordero et al. 2019) which, in turn, has a considerable impact on the geomorphological processes. Consequently, these systems are transgressive and with the potential to be highly dynamic. The aeolian landforms, conditioned by the vegetation and sediment supply, range from nebkhas to transverse ridges. Their high complexity, fragile nature, uniqueness and importance from the point of view of biodiversity and geodiversity explain why most of them have been granted protection status through different local, regional, national and international (EU) regulations.

These systems have evolved differently depending on their individual natural characteristics and the degree and type of anthropic pressure to which they have been subjected. Below, a joint analysis is presented of these aeolian sedimentary systems from the perspective of historical ecology. The systems considered in this study are the following: the coastal aeolian sedimentary systems of La Graciosa Island, El Jable (Lanzarote), El Médano (Tenerife), Jandía and Corralejo (Fuerteventura), and Maspalomas and Guanarteme (Gran Canaria) (Figures 19.1 and 19.2).

Figure 19.1. *Some of the sedimentary systems studied. Top-left: Jandía, top-right: Corralejo (by Claudio Moreno Medina), bottom-left: El Médano and bottom-right: Maspalomas. For a color version of this figure, see www.iste.co.uk/decocq/ecology.zip*

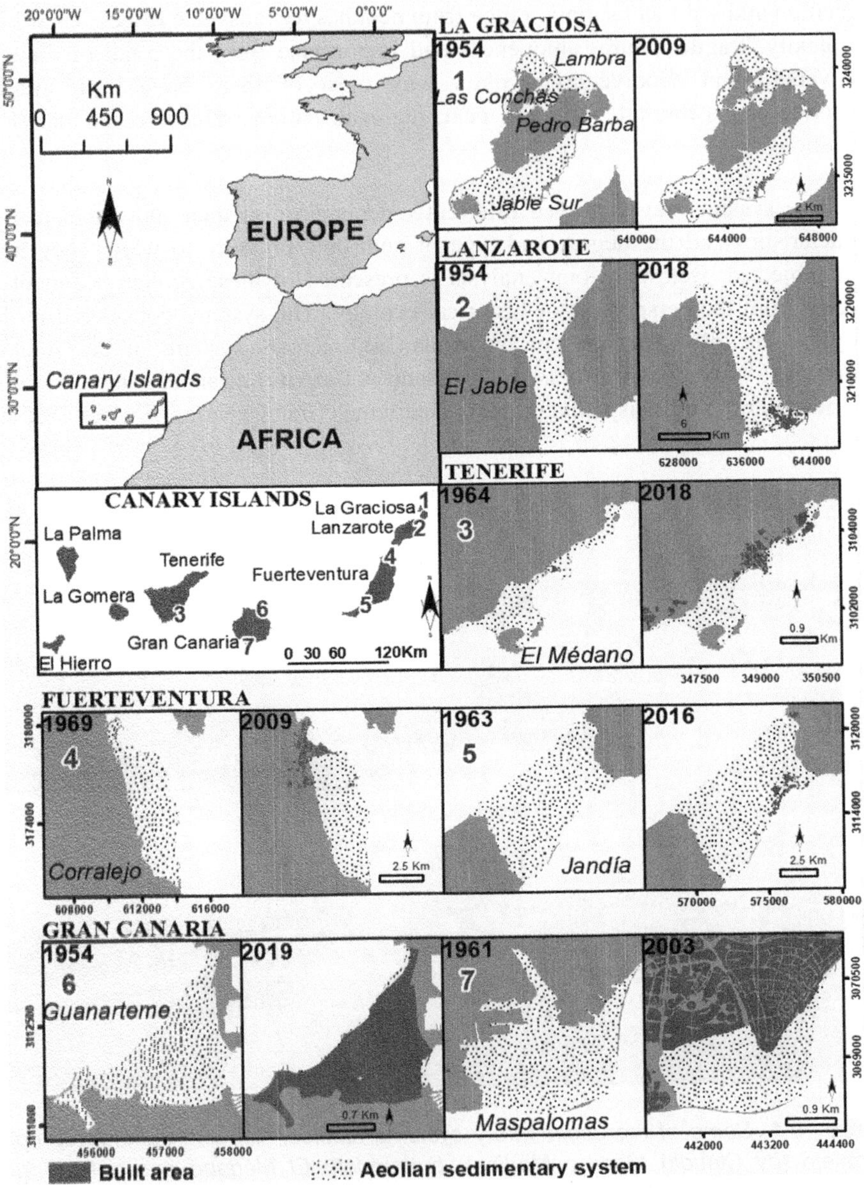

Figure 19.2. *Study sites. Old and current status of the arid coastal aeolian sedimentary systems studied. The time intervals considered correspond to those used in the studies conducted in García-Romero et al. (2016), Santana-Cordero et al. (2016a, 2016b) and Marrero-Rodríguez et al. (2020a, 2020b). For a color version of this figure, see www.iste.co.uk/decocq/ecology.zip*

19.3. Historical evolution of the coastal aeolian sedimentary systems of the Canary Islands

The anthropic impact on the aeolian sedimentary systems is directly related to the productive models that have been in place over the course of the islands' history since their conquest by the Spanish. This will therefore form the basis of the line of reasoning that will be followed in the presentation of the transformations that these systems have undergone due to changes in land-use.

In 1730–1736, the volcanic eruption of Timanfaya (Lanzarote) buried a significant amount of land used for agrarian purposes, including crop cultivation, grazing and firewood gathering. One result was the intensification of land-uses in El Jable (Lanzarote) and La Graciosa, while another was an increase in migration to Corralejo (Fuerteventura). After this eruption, herd grazing on La Graciosa became a permanent activity and was accompanied by the first attempts at the establishment of a population settlement. The consequent pressure that was imposed on the environment led to the issuing of local ordinances and regulations which, among other things, prohibited permanent settlements, the cutting or burning of shrubs, and the unauthorized entrance of livestock. A similar deforestation process was taking place in El Jable (Lanzarote) related, in this case, to the burning of plant material to manufacture liquors and the plowing of land for crop growing (Caballero-Mújica 1991). As a result of these processes, sediment remobilization took place through the action of aeolian sedimentary dynamics. In El Jable (Lanzarote), for example, there are descriptions of large mounds of sand crossing the island from north to south (Torriani 1959), while in Guanarteme (Gran Canaria), the historical cartography evidences the presence of landforms from 1599 and, over a century later, the change in orientation between 1742 and 1792 of a large dune in its northern sector (Santana-Cordero et al. 2014). In other systems, however, similar processes do not appear to have taken place, certainly not at any significant scale. For example, in Maspalomas (Gran Canaria), with some 100 inhabitants towards the end of the 18th century (Marrero and Naranjo 1999), the few land-uses that existed were non-intensive, and no significant environmental transformations seem to have been generated.

In general, the studied aeolian sedimentary systems were perceived socially from two distinct perspectives, which caused different anthropogenic pressures on them. On the one hand, some of the systems were seen as marginal territories due to their distance from the main population settlements and their aridity (lack of water resources) and consequent unsuitability for agriculture, especially in Gran Canaria and Tenerife (cases of Maspalomas and El Médano) which enjoyed considerably more abundant agroforestry resources. For example, the location of Maspalomas, at the southern tip of Gran Canaria and at some distance from the island's main population centers, accentuated its isolation and, possibly, the continued preservation of many of its natural features. On the other hand, in the dune systems of Lanzarote, Fuerteventura and La Graciosa, the general characteristics of aridity, low height and scant wood resources provoked the establishment of more intensive land-uses in their aeolian sedimentary systems (Marrero-Rodríguez et al. 2020a), as well as the introduction of more profound modifications. Exceptionally, Guanarteme was seen from both these perceptions during its history, i.e. as a marginal place first, and subsequently as an essential place for occupation.

19.3.1. *19th century: territorial consolidation and spread of the agrarian socioeconomic system*

Over the course of this period, the evolution of the aeolian sedimentary systems situated on islands with fewer agrarian resources (and where a relative overexploitation of these marginal areas took place) differed from the evolution of other systems located on islands with more resources. Another differentiating aspect was the development of the limestone industry around some of the systems and the consequent exploitation of the resources required for its practice. Indeed, this industry had perhaps one of the biggest impacts on the systems during this period, as not only did the limestone itself have to be quarried to provide the raw material (basically calcareous rock from coastal paleo-levels underlying loose aeolian sediments), but vegetation also had to be cut and gathered from the area to fire the limekilns. This deforestation led in turn to the remobilization of a significant volume of sediment and major changes to the landforms of these systems. The impact of this activity was especially notable in Jandía, Fuerteventura (Marrero-Rodríguez et al. 2020a). Together with extensive livestock grazing (mainly goats and sheep), this activity severely impacted the vegetation of the area. The effects of goat and sheep grazing were also significant in the aeolian system of Corralejo, in the north of the same island.

Another example of relative environmental overexploitation took place in La Graciosa, where there is evidence of a drastic reduction in plant cover (Santana-Cordero et al. 2016a) and generalized sediment remobilization processes. This was due not only to the use of limekilns, but also because of a notable population increase from 1880 onwards which added pressure on the environment through grazing, plowing, sand extraction for construction purposes, etc. In the case of El Jable (Lanzarote), a sand-based crop system was even established for which an extensive sector of the aeolian sedimentary system was prepared using plant-based windbreaks to slow down the circulation of sediment.

For its part, the aeolian sedimentary system of Guanarteme (Gran Canaria) had managed to retain many of its natural features despite bordering the island's capital, with 8.35–11.14 m high dunes being described in 1815 (von Buch 1999). By the mid-point of the century, however, several land-uses can be found in this system (Santana-Cordero et al. 2016b), including for agricultural, residential, recreational and extraction purposes. The consequences were modifications to a system that was perceived by the society of the time as an enemy that had to be conquered, as the sands were invading crops and road networks and approaching the entrance to the city. Contrastingly, in the southern area of the island, the Maspalomas dune system remained in a good state of conservation, principally because of the continued lack of interest in that space and, consequently, the low levels of anthropic pressure on the dune field (Marrero and Naranjo 1999).

19.3.2. *20th century to the present day: the tourism transformation*

The evolution of the different aeolian sedimentary systems considered in the present study varied substantially between the first half of the 20th century and the following decades when the urbanization process and the development of tourism became more widespread (Figure 19.1), albeit at differing rates and intensities.

In the first 50 years of the last century, there was a gradual decline in agrarian land-use in some of the aeolian sedimentary systems. At the same time, the import of fossil fuels began to reduce the pressure on vegetation, allowing plant recolonization and the reappearance of nebkhas in some islands, as occurred in El Jable (Lanzarote) and La Graciosa. However, in Jandía (Fuerteventura), oral sources suggest the opposite process was taking place in the period 1940–1960, with intensive deforestation taking place (Marrero-Rodríguez et al. 2020a). This can be attributed to a simultaneous increase in grazing and in the demand for firewood for the limestone industry. The aeolian sedimentary system was significantly destabilized and plant regeneration so affected that limestone began to be exported as raw material because there was insufficient plant material to fire the limekilns. At around the same time, intensive tomato growing became established between the 1930s and 1960s in Maspalomas (Gran Canaria), which meant the plowing and preparation of 80.9 ha situated in the northern sector of the dune field. This new activity additionally resulted in a population increase as people began to migrate from the north to the south of the island in search of work, with the number of inhabitants rising to 13,384.

However, the factor that by some distance most impacted the socioeconomic system of the islands was the emergence of tourism. A radical change in land-use took place from the 1960s onwards in Maspalomas (Gran Canaria), as well as a significant revaluation on the part of society of two resources which were key to the emergence of mass tourism, sun and beach, which had until then been virtually ignored. The development of the area as a tourist resort continued throughout the 1970s, 1980s and 1990s (Parreño Castellano 2001), with the area surrounding the dune field occupied by buildings and infrastructures of various kinds, as well as a golf course. This resulted in a major alteration to the aeolian sedimentary dynamics and, consequently, various ecological processes (Hernández-Calvento et al. 2014; Hernández-Cordero et al. 2017, 2018). In Corralejo (Fuerteventura), although the first bungalows were constructed in 1962, it was not until 1975 and the construction of two large hotels inside the aeolian sedimentary system that the tourist industry began to make its presence felt on the environment. Likewise, in El Jable (Lanzarote), a significant change in the economic model emerged, with traditional activities giving way to tourism and provoking a dramatic increase in anthropic pressure on the dune system. The first tourists began to arrive in El Médano (Tenerife) in 1964, and it was not long before cultivation of the tomato was gradually abandoned. Finally, La Graciosa also initiated an orientation of its economy towards mass tourism, with an almost doubling of the built-up area for its population center between 1977 and 2009 and a considerable extension of the island's port area.

Tourism entailed enormous changes to the economic and social structure, as well as to various landscape features (most notably in Maspalomas, Corralejo, Jandía, El Médano and La Graciosa). The construction of housing, tourist complexes and all manner of infrastructure resulted in the transformation of these systems and important environmental consequences (García-Romero et al. 2016; Table 19.1): (i) a decrease in the surface area of aeolian sedimentary systems (surface variation %), particularly in Maspalomas and El Médano, (ii) an increase in plant cover (vegetation variation %) and (iii) a decrease in the surface area occupied by bare sand (bare sand variation %).

In addition, a reduction of the distribution in the systems of some plant communities took place. For example, in Maspalomas, studies have been conducted on the reduction of the area covered by *Traganum moquinii* (-24.4%), essential for foredune formation (García-Romero et al. 2021), and *Plocama pendula* (-99.8%), as well as on the total disappearance of the *Euphorbia balsamifera* community (Hernández-Cordero et al. 2017). In other systems, such as Guanarteme, the total disappearance of *T. moquinii* communities has been confirmed (Santana-Cordero et al. 2014), while in La Graciosa, this plant community was severely impacted by its use as firewood for cooking (Santana-Cordero et al. 2016a). A dramatic reduction took place in Jandía in the number of specimens of *Convolvulus caput-medusae* because of its use as a fuel (Marrero-Rodríguez et al. 2020a), while the reduction of

T. moquinii in El Médano was due to the opening of extractive quarries which generated flooded areas where the plant was unable to grow (Marrero-Rodríguez et al. 2020b). An important coastal erosion also occurred in Maspalomas and Jandía, evidenced in the regression of the coastline (Alonso et al. 2002; Hernández-Cordero et al. 2018). Meanwhile, the Guanarteme system had been completely buried by the 1960s as a result of the incessant expansion of the city of Las Palmas de Gran Canaria (Santana-Cordero et al. 2014).

System/indicators	Surface variation (%)	Vegetation variation (%)	Bare sand variation (%)
Maspalomas (Gran Canaria) (1961–2003)	-24.0	38.8	-61.3
Corralejo (Fuerteventura) (1969–2003)	-12.9	8.9	-32.0
Lambra (La Graciosa) (1954–2009)	0.0	12.1	-47.1
Jable Sur (La Graciosa) (1954–2009)	-1.5	60.9	-77.5
Jandía (Fuerteventura) (1963–2016)	-3.6	50.1	-62.9
El Médano (Tenerife) (1964–2018)	-50.9	-18.5	-6.2

Table 19.1. *Main land cover changes over the past 70 years in the aeolian sedimentary systems of the Canary Islands. Modified from García-Romero et al. (2016) and Marrero-Rodríguez et al. (2020a, 2020b)*

Eventually, a growing awareness of the pressure being exerted on these systems through urban and tourism developments, as well as of the value and importance of the biodiversity and geodiversity of these ecosystems, led to them being declared protected nature reserves (Act 12/1987, dated June 19, on Natural Protected Areas in the Canary Islands), with the exception of a major part of the El Jable system and the Guanarteme system which had already been built over. This legislation aimed not only to preserve what remained of these ecosystems by preventing further construction in them but also provided regulations on traditional uses, giving rise to the natural regeneration of the vegetation and, to a certain degree, of the ecological processes in most of the systems (Table 19.1).

19.4. Conclusion

Over the course of the historical period considered in this study, different land-uses associated with the prevailing economic systems of the time have been developed in the aeolian sedimentary systems of the Canary Islands. The transition from an agrarian-based to a tourism-based system resulted in the decline of some uses and the emergence of other new ones. The traditional activities of an agrarian society dominated between the 15th century and the start of the 20th century. However, the uses that were made of most of the dune systems were generally marginal, except on the islands of Lanzarote and Fuerteventura given the limited resources that these systems contained for an agrarian society. In this regard, the case of Guanarteme was unique, impacted at an earlier stage by the growth and expansion of the city of Las Palmas de Gran Canaria.

Land-use changed radically in the second half of the 20th century as a result of urban development and the rise of tourism, with significant environmental changes taking place: (i) the disappearance of the Guanarteme dune system, (ii) a reduction in surface area in some of the aeolian sedimentary systems and (iii) ecological and geomorphological change, related to a decrease in the numbers of some plant communities, increased plant cover, sediment stabilization and the emergence of coastal erosion processes.

In summary, these ecosystems are no longer considered an "inconvenience" and of little interest for an agrarian society. Instead, they have become a fundamental resource for the growth in tourism and the development of the Canary economy over the course of the last 70 years.

19.5. References

Alonso, I., Alcántara-Carrió, J., Cabrera, L. (2002). Tourist resorts and their impact on beach erosion at Sotavento beaches, Fuerteventura, Spain. *J. Coastal Res.*, SI 36, 1–7.

von Buch, L. (1999). *Descripción física de las Islas Canarias*. Graficolor, La Laguna.

Caballero-Mújica, F. (ed.) (1991). *Compendio brebe y fasmosso, histórico y político, en que (se) contiene la cituazion, población, división, gobierno, produziones, fábricas y comercio que tiene la ysla de Lanzarote en el año de 1776*. Ayuntamiento de Teguise, Teguise-Las Palmas.

García-Romero, L., Hernández-Cordero, A.I., Fernández-Cabrera, E., Peña-Alonso, C., Hernández-Calvento, L., Pérez-Chacón, E. (2016). Urban-touristic impacts on the aeolian sedimentary systems of the Canary Islands: Conflict between development and conservation. *Isl. Stud. J.*, 11(1), 91–112.

García-Romero, L., Hernández-Cordero, A.I., Hesp, P.A., Hernández-Calvento, L., Santana del Pino, A. (2021). Decadal monitoring of Traganum moquinii's role on foredune morphology of an human impacted arid dunefield. *Sci. Total Environ.*, 758, 143802.

Hernández-Calvento, L., Jackson, D.W.T., Medina, R., Hernández-Cordero, A.I., Cruz, N., Requejo, S. (2014). Downwind effects on an arid dunefield from an evolving urbanised area. *Aeolian Res.*, 15, 301–309.

Hernández-Cordero, A.I., Hernández-Calvento, L., Pérez-Chacón Espino, E. (2017). Vegetation changes as an indicator of impact from tourist development in an arid transgressive coastal dune field. *Land Use Policy*, 64, 479–491.

Hernández-Cordero, A.I., Hernández-Calvento, L., Hesp, P.A., Pérez-Chacón, E. (2018). Geomorphological changes in an arid transgressive coastal dune field due to natural processes and human impacts. *Earth Surf. Proc. Land.*, 43(10), 2167–2180.

Hernández-Cordero, A.I., Peña-Alonso, C., Hernández-Calvento, L., Ferrer-Valero, N., Santana-Cordero, A.M., García-Romero, L., Pérez-Chacón Espino, E. (2019). Aeolian sedimentary systems of the Canary Islands. In *The Spanish Coastal Systems*, Morales, J.A. (ed.). Springer, Cham.

Marrero, J.L. and Naranjo, R. (1999). *Maspalomas espacio natural.* Ayuntamiento de San Bartolomé de Tirajana (Concejalía de Turismo), Las Palmas de Gran Canaria.

Marrero-Rodríguez, N., García-Romero, L., Sánchez-García, M.J., Hernández-Calvento, L., Pérez-Chacon Espino, E. (2020a). An historical ecological assessment of land-use evolution and observed landscape change in an arid aeolian sedimentary system. *Sci. Total Environ.*, 716, 137087.

Marrero-Rodríguez, N., García-Romero, L., Peña-Alonso, C., Hernández-Cordero, A. (2020b). Biogeomorphological responses of nebkhas to historical long-term land uses in an arid coastal aeolian sedimentary system. *Geomorphology*, 368, 107348.

Parreño Castellano, J.M. (2001). El proceso de urbanización del espacio turístico. In *Evolución e implicaciones del turismo en Maspalomas Costa Canaria (tomo I), vol. 1*, Hernández Luis, J.A. and Parreño Castellano, J.M. (eds). Ayuntamiento de San Bartolomé de Tirajana (Concejalía de Turismo), San Bartolomé de Tirajana.

Santana-Cordero, A.M., Monteiro-Quintana, M.L., Hernández-Calvento, L. (2014). Reconstructing the environmental conditions of extinct coastal dune systems using historical sources: The case of the Guanarteme dune field (Canary Islands, Spain). *J. Coast. Conserv.*, 18, 323–337.

Santana-Cordero, A.M., Monteiro Quintana, M.L., Hernández Calvento, L., Pérez-Chacón Espino, E., García-Romero, L. (2016a). Long-term human impacts on the coast of La Graciosa, Canary Islands. *Land Degrad. Dev.*, 27, 479–489.

Santana-Cordero, A.M., Monteiro-Quintana, M.L., Hernández-Calvento, L. (2016b). Reconstruction of the land uses that led to the termination of an arid coastal dune system: The case of the Guanarteme dune system (Canary Islands, Spain), 1834–2012. *Land Use Policy*, 55, 73–85.

Torriani, L. (1959). *Descripción e historia del reino de las Islas Canarias antes Afortunadas, con el parecer de sus fortificaciones (Vol. 2).* Goya Ediciones, Santa Cruz de Tenerife.

20

Historical Forest Microclimates

Emiel De Lombaerde[1], Karen De Pauw[1], Pallieter De Smedt[1], Jonathan Lenoir[2], Camille Meeussen[1], Thomas Vanneste[1], Kris Verheyen[1], Florian Zellweger[3] and Pieter De Frenne[1]

[1] *Ghent University, Belgium*

[2] *Jules Verne University of Picardie, Amiens, France*

[3] *Swiss Federal Institute for Forest, Snow and Landscape Research, Birmensdorf, Switzerland*

The majority of terrestrial organisms live in forests where microclimatic conditions below trees differ significantly from conditions outside forests. The temperature, humidity, wind and light environment in forest understories are strongly affected by the structure and composition of the forest, local topography, landscape characteristics and the regional macroclimatic conditions, among others, leading to highly variable microclimates in structurally contrasting forests and along successional or management gradients (Geiger et al. 2003). Temperature extremes are often strongly buffered in forests compared to open habitats, with cooler below – canopy maximum temperatures, warmer minimum temperatures and lower seasonal and interannual variability (De Frenne et al. 2021).

Climate warming, land-use change and the history of forest management are important determinants of the temporal dynamic of forest microclimates. However, very little is known about the history (and future) of forest microclimates, i.e. how microclimate conditions in the understory of forests are changing over time. The knowledge gap on the history of forest microclimates chiefly stems from a lack of long-term forest microclimate time series. Long-term climatic time series are almost exclusively based on weather stations located outside forests in open landscapes (with some exceptions such as the FLUXNET, ICP forests or rare regional networks). Given the lack of long-term time series of forest microclimates, current assessments of climate-change impacts on the biodiversity living in the shade of trees are thus subject to considerable uncertainty. Reconstructing the historical microclimatic conditions in forest understories is important to capture the realized climatic niche of forest species and to improve our understanding of the actual response of forest biodiversity to direct human pressures on forest ecosystems (e.g. disturbances due to forest management) as well as indirect human pressures through anthropogenic climate change. Time series of forest microclimate data are therefore key for assessing time-lag dynamics in forest communities, as well as for studying the effect of climate microrefugia on community responses to climate change.

In this chapter, we focus on historical microclimates and how they can help us to predict the future. We summarize the drivers and effects of past, present and future climate, land-use and forest management on temporal dynamics in understory microclimate (summarized in Table 20.1), methods to infer historical microclimates and outline the implications for forest biodiversity. The focus is on temperatures near the forest floor (0–1 m), because these temperatures are of high biological importance for understory plants and animals. The focus on temperature also provides a direct link to climate warming, although we clearly acknowledge that other microclimatic parameters, such as soil moisture and light availability, are also important.

Drivers of microclimate temperatures	Spatial scale	Importance for forest microclimates	Changing over ecological time?	Methods to infer the past
Tree species	Plot	+++	Yes	Legacy vegetation surveys, remote sensing, dendrochronology
Litter quantity and quality	Plot	+	Yes	Legacy vegetation and soil surveys
Forest density, structure	Plot	+++	Yes	Legacy vegetation surveys, remote sensing, dendrochronology
Tree height	Plot	+	Yes	Legacy vegetation surveys, remote sensing
Understory community composition	Plot	+	Yes	Legacy vegetation surveys, remote sensing
Edge effects	Patch	+++	Yes	Historical maps
Slope, aspect, topography	Patch (and landscape)	+++	No	NA

Drivers of microclimate temperatures	Spatial scale	Importance for forest microclimates	Changing over ecological time?	Methods to infer the past
Stand size and diversity, management system	Patch	+++	Yes	Historical maps, remote sensing, archival sources
Soil type, parent material	Landscape (and patch)	+	No	NA
Landscape composition	Landscape	++	Yes	Historical maps, remote sensing
Fragmentation per se vs. habitat area	Landscape	++	Yes	Historical maps, remote sensing
Macroclimate	Landscape	+++	Yes	Weather stations and downscaling, mechanistic models
Water bodies, rivers, lakes, sea	Landscape	++	No	NA

Table 20.1. *Overview of the drivers of forest microclimate from the plot to the landscape scale, their importance in affecting microclimate temperatures and whether the effect of the driving factor can change over ecological time*

20.1. Drivers of microclimate at the plot, forest and landscape scale

At a very local scale within a forest stand, the microclimate is modified by a complex interplay of different local factors: forest structure, tree species composition and height, the abundance of understory plants and topography. At this scale, the magnitude of temperature offsets (i.e. the horizontal difference between open and forest conditions) depends on episodic (e.g. due to silvicultural practices) or cyclic (e.g. due to the seasonal phenology) changes in forest canopy density and composition, tree height and understory plants. Tree structure and composition affect wind speeds and patterns, incoming light quantity and quality, moisture availability and finally also sub-canopy and soil temperatures (Geiger et al. 2003). Even historical disturbances, for example, due to temporary canopy openings, are known to affect the understory communities and cause a shift towards warm-adapted species in forest-floor plant communities. Tree species composition can directly, and indirectly via the forest structure, affect understory microclimates. Species differ in tree and crown geometry and leaf area index, and therefore also in their ability to cast shade. Hence, deep shade casting trees such as beech (*Fagus sylvatica*) have a large crown depth, and consequently low understory light levels. Differences also occur between broadleaved and coniferous forests, with broadleaved forests showing higher evaporative cooling and a higher ability to buffer maxima temperatures after canopy closure. Historical forest transformation of broadleaved forest and afforestation with conifers have therefore had a considerable impact on forest microclimates. Moreover, trees and shrubs also affect the soil microclimate via their litter quantity and quality. Thick litter layers, as found under species with a low litter quality (e.g. beech), further reduce soil temperature variability and buffer maxima and minima temperatures while reducing soil evaporation. Finally, also the understory community composition might further influence the forest floor microclimate via similar mechanisms as for the litter layer, in addition to regulating below-tree canopy microclimates via transpiration.

At the forest stand scale, the management system and edges (see the next paragraph) are the most important drivers determining spatial and temporal variation in microclimates (Table 20.1). First, there is a major difference in microclimate between coppice and coppice-with-standards (CWS) systems, on the one hand, and high forest-systems, on the other hand. Coppice and CWS systems are characterized by clear cuts of the coppice layer every 10 to 20 or 30 years, depending on the species and/or the management target. These frequent stand-level cuttings have a strong impact on the microclimate and lead to repeated cycles of low microclimatic buffering being replaced by periods with high buffering. Similar, but more tempered, patterns develop in CWS systems where a sparse canopy of standard trees remains in place all the time. Different microclimatic regimes are encountered in high forest systems (Figure 20.1), due to the substantially longer rotation times in between regeneration cuts, ranging from multiple decades to centuries, and due to the generally smaller size of the regeneration cuts. Stand-scale cuts that strongly disrupt the microclimatic buffering do occur in clear-cut and shelterwood systems. However, the frequency of these cuts is much lower than in coppice and CWS systems. The regeneration cuts in strip cutting, group selection and single-tree selection systems occur on a smaller spatial scale (Figure 20.1), so that the microclimatic buffering is more or less maintained. Currently, the vast majority of forests is managed as high forest, but coppice-based systems were in many places dominant until the middle of the 20th century. This means that forest microclimates have undergone a very large shift over the course of the last century. More strongly buffered forest microclimates are now much more prominent than in the past (e.g. Zellweger et al. 2020). This microclimatic shift is a major driver for the large shifts in forest biodiversity that is observed in many places throughout Europe (see section 20.3). Finally, it should be stressed that the effects of the applied management system extend beyond the stands where the regeneration cuts take place. Systems where large cuts are applied (notably coppice, CWS, clear-cut and shelterwood systems) also generate substantial internal forest edges that cause important microclimatic changes in adjacent stands.

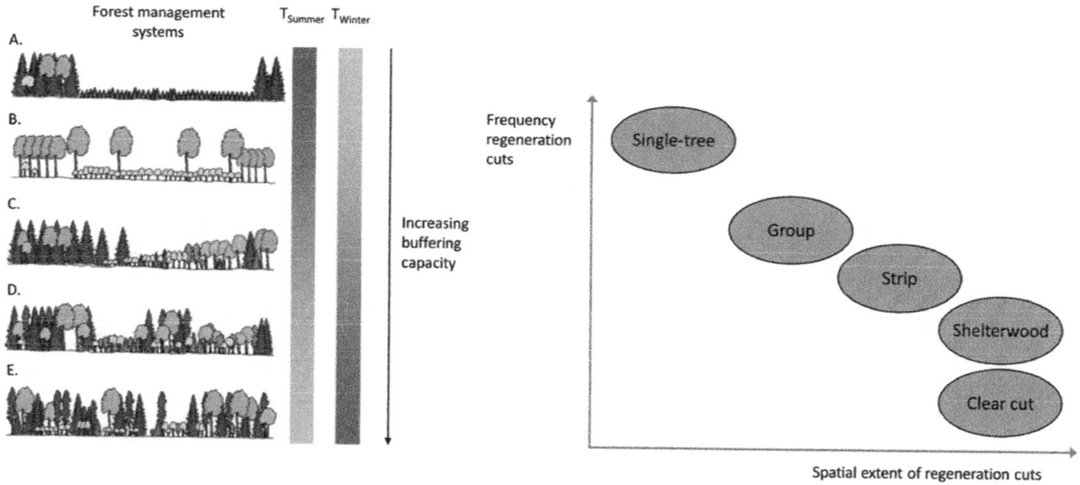

Figure 20.1. *Left panel: schematic representation of five common high-forest management systems (A: clear cut, B: shelterwood system, C: strip cutting system, D: group selection system and E: single-tree selection system). The buffering capacity of the forest canopy (black arrow) and resulting microclimates will vary between management systems. For instance, summer temperatures (Tsummer) will be warmer in forest that have undergone large-scale canopy removal such as in clear cuts (A) compared to forests where canopy disturbances are limited (E); for winter temperatures (Twinter), the opposite will be the case. Right panel: the spatial and temporal scale of the regeneration cuts applied in the five management systems. Figures adapted from den Ouden et al. (2010). For a color version of this figure, see www.iste.co.uk/decocq/ecology.zip*

Across the world, forests are becoming smaller due to habitat loss and fragmentation. The high degree of forest fragmentation results in forest edges increasingly becoming a dominant landscape feature. Worldwide, 20% of forest area lies within 100 m of a forest edge, but edge effects can penetrate much deeper into forest patches depending on the forest type. Forest edges have a very contrasting microclimate compared to forest interiors with, for example, higher air and soil temperatures, lower soil moisture, higher light availability and higher wind speed compared to forest interiors. These microclimatic conditions are induced by the exposed character of forest edges, and also by their altered vegetation structure. As a result, forest edges harbor distinct communities of plants, vertebrates, arthropods and microbes. Forest microclimates are increasingly altered through edge effects since forest fragmentation and hence edge creation is a still ongoing process. Based on forest edge research, we can assume that microclimate temperatures in forest patches have increased in the past due to the increasing proportion of edges compared to interior forest.

The forest microclimate is not only affected by plot and stand characteristics, but also by the landscape in which it is embedded. Several characteristics of the surrounding landscape influence the forest microclimate, either in a direct or indirect way. Topography is one of the most determining factors of a landscape and can have a strong impact on the microclimate, but does not change strongly over time. The same applies to the presence and size of large water bodies. However, other landscape features are of a dynamic nature and these are of prime importance to estimate how historical (and future) microclimates at the landscape-scale differ from today (Table 20.1). Microclimatic buffering is related to the macroclimate, a feature changing on an even larger scale than the landscape scale. Increasing ambient air temperatures due to macroclimate change might lead to increased buffering of maximum temperatures in forests. Furthermore, regional hydrological conditions and varying levels of soil moisture can affect microclimatic buffering through the evapotranspiration and the transformation of sensible to latent heat. Higher soil moisture reduces diurnal temperature variation and warming due to incoming radiation. As a result, also future changes in local water balances are important: forests might lose their microclimatic buffering capacity if they become water-limited. In addition to climate change, land-use change can affect forest microclimates. For example, urbanization and urban centers induce an urban heat island. Temperatures are typically higher in urbanized areas. Minimum forest temperatures are warmer in urban forests. The presence of urban areas has changed strongly over time and is still growing rapidly, as the global distribution of impervious surface area more than doubled since 1990. Similarly, temporal changes in the landscape configuration, for example, configuration of forest patches within the landscape, in terms of spatio-temporal continuity and connectivity, can potentially affect the history of local climatic conditions. However, knowledge on the influence of landscape composition and configuration on forest microclimates is still limited and should be extended, as landscapes have changed due to anthropogenic and climate-change drivers and will keep on changing in the future.

20.2. Methods to infer microclimate from the past and predict into the future

To infer microclimates from the past at a plot scale, we need legacy data on structural variables and tree composition, height and understory plants. These can chiefly be derived from: (i) historical surveys, (ii) chronosequences, (iii) remote sensing data (e.g. back to the 1980s using Landsat imagery) and (iv) dendrochronological data. In that respect, the legacy of botanists who collected numerous vegetation plots now stored in vegetation databases (Chytrý et al. 2016; Verheyen et al. 2017) represents a wealth of information and a unique opportunity to infer past microclimatic conditions in forest understories and thus potentially for reconstructing long-term time series of forest management and microclimate (Zellweger et al. 2020). As historical data on forest microclimates are scarce and difficult to gather, time series of detailed data on local forest structure and composition collected within national forest inventories as well as other vegetation resurvey programs provide unique opportunities to shed light on how forest microclimates have changed over the past century. One promising route is to link the horizontal (i.e. keeping the height above ground fixed) offset between current macroclimate and microclimate conditions to current plot or stand-level forest attributes, such as canopy cover and canopy composition, and apply the resulting relationships to the same forest attributes measured in past field surveys, thus reconstructing the forest microclimate of the past (Zellweger et al. 2020). Similarly, time series describing changes in forest structure and composition provide important input parameters for modeling microclimates over time based on mechanistic processes governing the small-scale variation in temperatures (Maclean et al. 2019). Likewise, chronosequences and "structural sequences" from young forests on post-agricultural land, older post-agricultural forests and ancient forests with contrasting tree ages and structures can, via a space-for-time substitution approach, allow ecologists to infer microclimates from early- versus late-successional stages. Forest structural attributes can be obtained from baseline and resurveys of (quasi-)permanent plots and also from repeated remote sensing (see further). A third approach is to use dated tree-ring data to reconstruct forest structure and microclimate. Indeed, the reconstruction of past forest structure and management is possible via analysis of tree rings and growth time series; for instance, approximate dates of management conversion, tree felling or extreme droughts can be interpolated from wood cores. Finally, dynamic forest growth models might be useful in this context too. Such models are often used for forecasting into the future (e.g. in relation to global change) albeit hindcasting with the goal of inferring past microclimates should be possible too but has been rarely done.

Historical maps to document past forest cover allow us to go 150–250 years back in time in many places across Europe and are crucial in estimating past forest microclimates. Very old maps, such as the French Cassini maps and the Belgian Ferraris map, both of the 18th century, can be used to date forest patches and thus distinguish between ancient and more recent forest patches harboring very different levels of forest biodiversity and ecosystem services and thus potentially different forest microclimates. Historical microclimates of forest patches can thus be deduced from old forest maps, by calculating forest edge length. Forest type-specific and region-specific gradients of microclimatic variables like soil temperature (see, for example, Arroyo-Rodríguez et al. 2017) and soil moisture can be used to estimate microclimatic conditions at forest patch level from the past, at least in small forest fragments with a high edge-to-interior ratio. Unlike forest edges, historical management systems are less easy to deduce from historical maps. Typically, a variety of sources have to be used to reconstruct past management. When available, archival documents about past management, exploitations and wood sales are the most obvious source of information, and can be used in conjunction with other sources such as pollen profiles to reconstruct tree species composition (Szabó 2015). However, written records are often lacking and, in those cases, we have to rely on field observations, comprising the composition and structure of the tree layer, the shape of trees and their growth patterns, assessed via dendrochronological methods. In large forest fragments, species composition and management are likely to fulfill a larger role in predicting historical forest microclimatic conditions compared to forest edge. In addition, forest management and tree species composition are main drivers in edge structural diversity. Microclimatic gradients in forest edges are, however, also dependent on the surrounding matrix stressing the importance of taking the landscape level into account when assessing historical microclimates at forest patch level.

Mapping and monitoring temporal microclimatic variation at the landscape scale is key to understanding organismal responses to climate variation, which in turn helps to predict how climate change will impact biodiversity and ecosystems. Yet, former approaches to map landscape-level microclimates were bound to several limitations. For instance, networks of microclimatic sensors provide point-based measurements while weather stations deliver macroclimatic data, but methods to effectively interpolate or downscale this data have been lacking so far for forests (Maclean et al. 2019). Several new techniques to map and model microclimates have emerged over recent decades. Probably one of the most promising techniques is remote sensing, producing exceptionally detailed and spatially continuous data layers that can feed into models of horizontal and vertical microclimatic variation of broad spatio-temporal scales (Zellweger et al. 2019). Remote sensing data can also be used to infer microclimates at the plot scale (see above).

First, a relatively simple technique to assess microclimates at the landscape level is the use of land-use maps that offer, sometimes very detailed, snapshots of the landscape configuration including the position, size and connectivity of forest patches at the time they were drawn (see above). Digitizing a chronosequence of historical land-use maps from different points in time may be used to quantify spatio-temporal changes in a variety of landscape attributes including the amount of forest cover and forest continuity. Beyond land-use maps, airborne and spaceborne (e.g. NASA Global Ecosystem Dynamics Investigation; GEDI) light detection and ranging or "laser scanning" (LiDAR) is now increasingly applied in ecological studies for a 3D analysis of the Earth's surface and vegetation (Zellweger et al. 2019). LiDAR data can then be combined with sensor networks providing in situ measurements of microclimatic conditions on the ground, using statistical approaches to predict microclimates across the entire LiDAR-mapped area (Frey et al. 2016a). This technique has been proven effective to model radiation regimes along vertical canopy profiles in forests as well as light transmittance to the forest understory at any time in the day or year (Tymen et al. 2017).

Alternatively, aerial photography can be used to map ground topography and vegetation structure via photogrammetry or structure-for-motion techniques. Thermal imaging with thermal infrared (TIR) cameras can also be applied to measure surface temperatures across large spatial scales. TIR cameras can also be installed on unmanned aerial vehicles allowing for a landscape-level mapping at a centimeter resolution. Accurate surface temperature measurements with TIR imagery are to date still challenging because surface temperatures may still deviate from actual microclimatic air temperatures experienced by organisms or because not all surfaces have a similar thermal emissivity.

Finally, another approach to generate microclimate maps is to downscale macroclimatic grids such as WorldClim (Fick and Hijmans 2017) with high-resolution remote sensing data layers such as digital terrain models or canopy height models. An important advantage of this technique is that these models are based on long-term series of macroclimatic data, thus allowing predictions of how microclimatic conditions vary across time (Zellweger et al. 2019). This is particularly relevant for forests where understory plant community changes are primarily driven by long-term microclimatic dynamics rather than the rate of macroclimatic warming (Zellweger et al. 2020). Mechanistic microclimate models offer an alternative method to downscale macroclimatic data based on a set of abiotic variables such as longwave and shortwave radiation, air temperature, rainfall, wind speed, relative humidity, elevation, coastal effects and canopy shading (Maclean et al. 2019).

20.3. Why do historical microclimates matter? Impacts on biodiversity from the plot to landscape scale

When plot data are collected from surveys that record concurrent forest microclimatic conditions and understory species composition (not only of plants, but also, for example, arthropods, fungi and soil bacteria) at the exact same location and period, impacts of local forest microclimate change can be linked to population, community and ecosystem change. For instance, species-specific response curves along a given microclimatic parameter (e.g. thermal response curves) can be fitted to compute indices of community weighted means, such as floristic temperatures, and, when expressed over time, calculate the increasing dominance of warm-adapted species, i.e. thermophilization. This provides interesting new avenues for ecological assessments based on floristic temperatures. Finally, also herbarium specimens with accurate collection coordinates offer novel opportunities to quantify impacts of global change on understory plants (e.g. on phenology), if they can be linked to historical microclimate data.

A study by Gorissen et al. (2004), which investigated the decline in woodland butterflies in Flanders (Belgium), illustrates the value of creating more thermal hotspots in forest stands. They used historical butterfly data from two woodlands in Belgium to relate the changes in butterfly diversity to changes in management practices. They found that almost all typical woodland butterflies and also species that make use of open habitats within woodlands went extinct in the second half of the 20th century. They linked these declines to afforestation of open areas and management change towards high forest practices. In the past, these forests were managed as coppice or CWS which allowed more light to enter the forests. They therefore argued that creating more open conditions such as larger ridges, more open area, may benefit woodland butterflies, and also thermophilous taxa from other groups. However, creating more open habitat may result in reduced temperature buffering capacity of tree canopies. The cool and moist microclimate that closed canopies induce is essential for many forest organisms adapted to lower temperatures, moist and shaded conditions, such as slugs, ectomycorrhizal and wood-decaying fungi, mosses and other cool-adapted plants, moths and specific groups of arthropods (fungivorous beetles, woodlice, etc.) (Figure 20.2). Many forests in Europe have become denser and cooler due to an overall reduction in forest management intensity during the past decades, which is beneficial for these forest specialist organisms. However, if forest cover will be reduced or canopies become more open in the future due to changes in disturbance or management, these organisms are likely to be negatively affected by increased edge effects.

Studying microclimates at the landscape scale can significantly contribute to understanding biodiversity and species distribution patterns. Landscape factors, such as forest cover and urbanization, can mediate the responses of wintering birds to winter temperatures through microclimate variability, access to resources or both (Frey et al. 2016b). Microclimate temperatures can have a large effect on the occupancy dynamics of forest birds in a montane forest landscape with high microclimatic variability. This suggests that the microclimate-associated distribution patterns reflect the species' potential for behavioral buffering from climate change. Also, for plants, diversity and distribution patterns can be linked to microclimatic differences and connectivity at the landscape scale. Local habitat features of hedgerows, including the microclimate, can shape plant diversity and composition patterns, affecting the potential of hedgerows to function as refugium or corridor for forest plant specialists (Vanneste et al. 2020). The increasing possibilities to map and model forest microclimates and its drivers at the landscape scale will provide species distribution models and biodiversity studies with important information to gain more insight in historical species' distributions and future developments under different global change scenarios.

Figure 20.2. *The butterfly Apatura iris (left) and the woodlice Ligidium hypnorum (right) as representatives of, respectively, hot- and cold-spots in forests. Photos: Wiske Teugels and Spinicornis-Gert Arijs. For a color version of this figure, see www.iste.co.uk/decocq/ecology.zip*

20.4. Conclusion

Forest microclimates have changed during the past century at the plot, forest and landscape level. These changes are still ongoing due to drivers such as climate warming, changes in forest management and land-use. In many forests, the degree of microclimatic buffering has increased due to a change from coppice-based systems to high forest. But at the same time, buffering along (external) edges has decreased over time due to increasing fragmentation, affecting the forest interior microclimate. To cater to the needs of forest organisms with contrasting microclimatic requirements in a changing climate, several management options are available. For instance, specific zoning systems can be developed to create both hot and cold spots in forested landscapes, considering the surrounding landscape matrix. In this endeavor, the historical knowledge of microclimates provides an invaluable source of information to build on.

20.5. References

Arroyo-Rodríguez, V., Saldana-Vazquez, R.A., Fahrig, L., Santos, B.A. (2017). Does forest fragmentation cause an increase in forest temperature? *Ecol. Res.*, 32(1), 81–88.

Chytrý, M., Hennekens, S.M., Jiménez-Alfaro, B., Knollová, I., Dengler, J., Jansen, F., Landucci, F., Schaminée, J.H.J., Aćić, S., Agrillo, E. et al. (2016). European Vegetation Archive (EVA): An integrated database of European vegetation plots. *Appl. Veg. Sci.*, 19(1), 173–180.

De Frenne, P., Zellweger, F., Rodríguez-Sánchez, F., Scheffers, B.R., Hylander, K., Luoto, M., Vellend, M., Verheyen, K., Lenoir, J. (2019). Global buffering of temperatures under forest canopies. *Nat. Ecol. Evol.*, 3, 744–749.

De Frenne, P., Lenoir, J., Luoto, M., Scheffers, B.R., Zellweger, F., Aalto, J., Ashcroft, M.B., Christiansen, D.M., Decocq, G., De Pauw, K. et al. (2021). Forest microclimates and climate change: Importance, drivers and future research agenda. *Glob. Chang. Biol.*, 27(11), 2279–2297

Fick, S.E. and Hijmans, R.J. (2017). Worldclim 2: New 1-km spatial resolution climate surfaces for global land areas. *Int. J. Climatol.*, 37(12), 4302–4315.

Frey, S.J.K., Hadley, A.S., Johnson, S.L., Schulze, M., Jones, J.A., Betts, M.G. (2016a). Spatial models reveal the microclimatic buffering capacity of old-growth forests. *Sci. Adv.*, 2, e1501392.

Frey, S.J.K., Hadley, A.S., Betts, M.G. (2016b). Microclimate predicts within-season distribution dynamics of montane forest birds. *Divers. Distrib.*, 22, 944–959.

Geiger, R., Aron, R.H., Todhunter, P. (2003). *The Climate Near the Ground*. Rowman and Littlefield, Oxford, UK.

Gorissen, D., Merckx, T., Vercoutere, B., Maes, D. (2004). Veranderd bosgebruik en dagvlinders. *Landschap*, 21, 85–95.

Maclean, I.M.D., Mosedale, J.R., Bennie, J.J. (2019). Microclima: An R package for modelling meso- and microclimate. *Methods Ecol. Evol.*, 10(2), 280–290.

den Ouden, J., Muys, B., Mohren, F., Verheyen, K. (2010). *Bosecologie en Bosbeheer*. Acco, Leuven, Belgium.

Szabó, P. (2015). Historical ecology: Past, present and future. *Biol. Rev.*, 90, 997–1014.

Tymen, B., Vincent, G., Courtois, E.A., Heurtebize, J., Dauzat, J., Marechaux, I., Chave, J. (2017). Quantifying micro-environmental variation in tropical rainforest understory at landscape scale by combining airborne LiDAR scanning and a sensor network. *Ann. For. Sci.*, 74, 32.

Vanneste, T., Govaert, S., Spicher, F., Brunet, J., Cousins, S.A.O., Decocq, G., Diekmann, M., Graae, B.J., Hedwall, P., Kapás, R.E. et al. (2020). Contrasting microclimates among hedgerows and woodlands across temperate Europe. *Agric. For. Meteorol.*, 281, 107818.

Verheyen, K., De Frenne, P., Baeten, L., Waller, D.M., Hédl, R., Perring, M.P., Blondeel, H., Brunet, J., Chudomelová, M., Decocq, G. et al. (2017). Combining biodiversity resurveys across regions to advance global change research. *BioScience*, 67(1), 73–83.

Zellweger, F., De Frenne, P., Lenoir, J., Rocchini, D., Coomes. D. (2019). Advances in microclimate ecology arising from remote sensing. *Trends Ecol. Evol.*, 34, 327–341.

Zellweger, F., De Frenne, P., Lenoir, J., Vangansbeke, P., Verheyen, K. et al. (2020). Forest microclimate dynamics drive plant responses to warming. *Science*, 368(6492), 772–775.

21

Causes and Consequences of Extinction Debts: Perspectives for Historical Ecology and Biological Conservation

Grégoire BLANCHARD[1] and François MUNOZ[2]

[1] *University of Montpellier, France*

[2] *Grenoble Alpes University, Grenoble, France*

21.1. Introduction

Among the concepts that are pivotal in historical ecology, extinction debt has received considerable attention and strong support from both theoretical and empirical evidence (Malanson 2008; Figueiredo et al. 2019). A broad definition of extinction debt is that, after a modification of ecosystem characteristics, we can observe populations of some species at a given time and place although they are doomed to become extinct under current ecological conditions (Kuussaari et al. 2009). Such extinction delay can last over decades or centuries (Hylander and Ehrlén 2013). Assuming that today's biodiversity patterns represent a snapshot of ecosystems at equilibrium can then be misleading for conservation planning, making uncertainty on the causes and consequences of these long transient dynamics a crucial concern for biodiversity research (Kuussaari et al. 2009). While anthropogenic alteration of ecosystems becomes faster and stronger, there is a growing need to address the diversity of ecological – and historical – scenarios leading to extinction debts, as well as implications and opportunities for biological conservation (Figueiredo et al. 2019).

Historical Ecology,

Accumulating long-term, empirical data on ecosystem history sheds light on the mechanisms generating delayed extinctions (Helm et al. 2005; Vellend et al. 2006; Vellend et al. 2013). An illustrative example of the usefulness of historical ecology in understanding the causes and processes generating extinction debt is the history of North American National Parks. At the end of the 19th century, several large nature reserves were settled in western North America. Despite strict conservation policies, a critical decline of mammal populations was observed within parks during the 20th century (Newmark 1995). Habitat fragmentation due to habitat destruction outside the Parks entailed the progressive loss of biodiversity, consistent with the dynamics predicted by the theory of island biogeography (MacArthur and Wilson 1967; Newmark 1995). Specifically, if protected areas are too small for long-term persistence of local populations and without immigration from outside the parks, a progressive decline of biodiversity is expected (Newmark 1995). Similarly, there is growing evidence for long time lags between expected and observed plant extinctions following anthropogenic landscape changes (Vellend et al. 2006; Cronk 2016). Historical ecology thus provides valuable opportunities for studying extinction debts and underlying processes, through taking advantage of long-term and large-scale empirical scenarios from the past.

In this chapter, we expose ecological mechanisms leading to extinction debts, and address implications for historical ecology from both research and conservation perspectives. We discuss how, by shedding light on past ecological dynamics, historical ecology is pivotal to detect and understand the mechanisms causing extinction debts. Hindcasting past extinction debts through the prism of ecosystem history can provide new insights and opportunities for future biodiversity conservation.

21.2. Causes and processes entailing extinction debts

Extinction debts arise from different mechanisms and under diverse historical scenarios (Hylander and Ehrlén 2013). Here, we propose a brief synthesis of the historical causes and ecological processes underlying extinction debts, showing how changes in basic properties of habitat (quantity, connectivity and quality) and of biotic interactions can trigger delayed extinctions (Figure 21.1).

The persistence of species populations depends on local environmental conditions, which represent the "habitat quality" for each species. Drastic alteration of habitat quality can lead to rapid population extinction, but smaller or more gradual changes can entail slow population decline (Hylander and Ehrlén 2013). Historical changes in habitat quality can generate delayed extinctions in areas progressively becoming unsuitable for population persistence. For instance, human-driven climate warming since the mid 20th century causes delayed plant species extinction in European Alps (Rumpf et al. 2019). Changing disturbance regimes (e.g. land-use practices, wildfires) can also affect habitat quality and result in extinction debts (Essl et al. 2015). Importantly, long-lived species can be subject to lags in demographic decline, notably if environmental change affects only specific life stages (e.g. reproduction or recruitment, Hylander and Ehrlén (2013)). For instance, individual trees may persist for hundreds of years even though low habitat quality prevents seedling recruitment (Vellend et al. 2006). Conversely, changes in environmental conditions can also generate an immigration credit for species adapted to novel environmental conditions (Box 21.1).

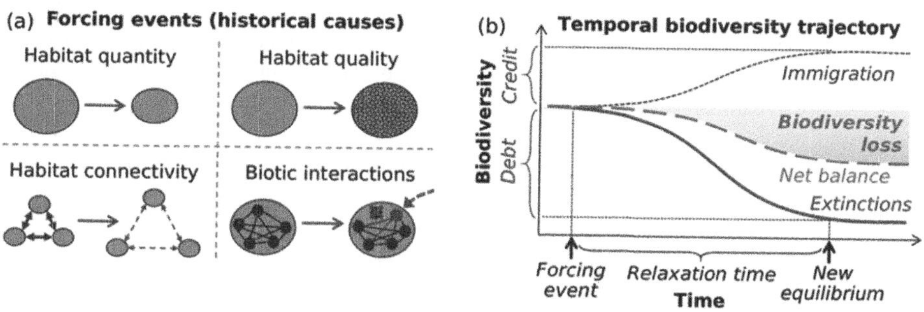

Figure 21.1. *Forcing events (historical causes) and temporal biodiversity trajectory associated with an extinction debt. (a) Red arrows indicate changes in abiotic (habitat characteristics) and biotic (species extinction or invasion) ecosystem conditions, which entail extinction debts. (b) Extinction debt is progressively paid during the relaxation time following the forcing events in (a), until a new equilibrium is reached. New environmental conditions in (a) can allow establishment of new immigrating species, but the immigration can also be delayed, resulting in an immigration credit (Box 1). Biodiversity change (grey area) depends on the net balance (grey dotted line) between extinctions (red line) and immigration (blue dotted line). For a color version of this figure, see www.iste.co.uk/decocq/ecology.zip*

Habitat loss and fragmentation are further drivers of extinction debt, which received most interest (Tilman et al. 1994). The available quantity of a given habitat directly determines the quantity of resources for species, and hence, the size of populations (MacArthur and Wilson 1967). Following past habitat loss, smaller remnant populations entail greater demographic stochasticity, increasing the risk of extinctions (Hylander and Ehrlén 2013). Habitat connectivity determines the ability of species to immigrate in habitat patches. Thus, lower connectivity resulting from past habitat fragmentation reduces immigration rates, and increases the probability of population extinction (Essl et al. 2015). Historical decreases in habitat quantity and/or connectivity are likely to push multiple species toward stochastic extinction, triggering large and long-lasting extinction debts (Hylander and Ehrlén 2013).

> At an equilibrium of extinction and immigration dynamics, biodiversity is expected to reach stationarity. Extinction debt is a transient biodiversity state during which species loss is delayed after a forcing event altering the initial ecosystem conditions (Figure 21.1). Conversely, the changes in abiotic or biotic ecosystem conditions can create opportunities for other species to immigrate. Because it can take time for these species to colonize new areas, immigration can be delayed as extinction can be, entailing an "immigration credit" (Jackson and Sax 2010). The "net" biodiversity change (i.e. loss or gain of species) depends on the difference between immigration credit and extinction debt (Figure 21.1). Immigration credit can be paid faster or slower than extinction debt is, generating transient biodiversity excess or deficit, respectively. The relaxation time represents the amount of time during which extinction debt and/or immigration credit are paid until biodiversity dynamics reaches the new equilibrium (Kuussaari et al. 2009).

Box 21.1. *Extinction debt, immigration credit and relaxation time*

Changes in biotic ecosystem context can also cause extinction debts (Mouquet et al. 2011; Figueiredo et al. 2019). For instance, the invasion by generalist and strong competitors, or by a new predator, can entail gradual extinction of native species (Didham et al. 2007; Gilbert and Levine 2013). These are classical scenarios on oceanic islands where many human-driven species invasions have generated extinction debts lasting for decades or centuries (Sax and Gaines 2008; Cronk 2016). Moreover, the extinction of a given species can entail the delayed co-extinction of other species through loss and disruption of interactions, generating cascading extinction debt (Valiente-Banuet et al. 2015). Delayed co-extinctions are particularly expected to arise following the extinction of keystone or engineer species having key roles in ecosystem functioning (Estes et al. 2011). A striking example of such delayed co-extinction dynamics is the slow decline of large-seeded tree species resulting from long-term human-driven defaunation of large-bodied dispersers in tropical forests (Terborgh et al. 2008).

21.3. Studying and detecting extinction debts from ecosystem history

Research challenges arising from extinction debts stand in the long-term delay between their historical causes and ecological consequences (Kuussaari et al. 2009). Extinction debts cannot be studied from snapshots of current ecosystem conditions, and often lag over timescales too long to match the duration of most ecological monitoring. Available data from the past, as provided by a variety of sources (see Vellend et al. 2013), are thus essential to investigate the trajectory of ecosystem dynamics and deduce the legacies of past ecosystem conditions and historical changes (Kuussaari et al. 2009; Figueiredo et al. 2019). Historical ecology is thus at the forefront of establishing causal links between past events and delayed extinction dynamics (Figure 21.2).

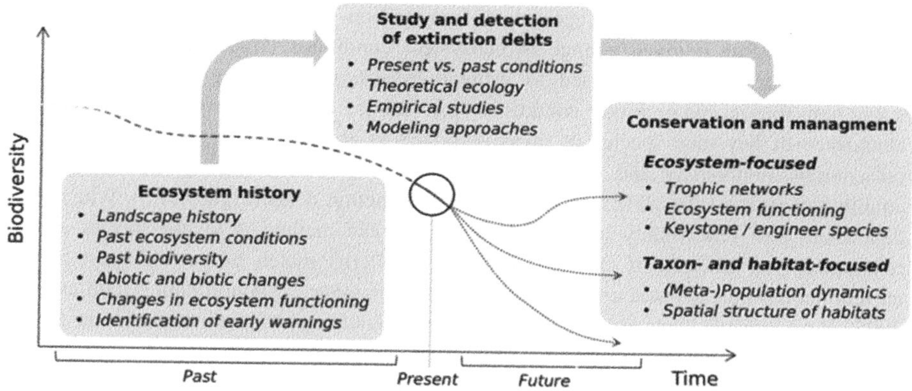

Figure 21.2. *Synthetic framework for studying, detecting and mitigating extinction debts. Information about ecosystem history helps to detect extinction debt and understand underlying mechanisms. Then, adequate conservation and management measures can be developed to mitigate extinction debts. Conservation perspectives focusing on ecosystem functioning should better succeed in maintaining or restoring ecosystem biodiversity and services than taxon- and habitat-focused approaches. For a color version of this figure, see www.iste.co.uk/decocq/ecology.zip*

Landscape ecology probably encompasses most of the empirical approaches used to assess delayed extinction dynamics (Box 21.2, Figure 21.3). Landscape history can be assessed through terrestrial and aerial historical photographs, as well as from historical maps (Vellend et al. 2013). Even when they are not related to a change in land-cover, detailed information about past land-use practices (e.g. agricultural and pastoral practices, soil fertilization, logging and forest management practices, hunting, stream water use) can also provide important insights about past habitat characteristics and ecosystem legacies (Vellend et al. 2013; Bürgi et al. 2017; Bergès and Dupouey 2020). Although only available for the last few decades, satellite images are also a promising source of information about landscape history for investigating extinction debts at the regional scale (Jung et al. 2019). Analyzing how species with different functional traits respond to recent changes in ecosystem conditions should help to identify the specific factors of extinction debts (Hylander and Ehrlén 2013). Such approaches can be used to infer parameters of mechanistic community assembly models (Blanchard et al. 2020), providing a more explicit understanding of the ecological mechanisms at play, and enabling simulations of future ecosystem dynamics (Box 21.2).

Detecting extinction debts can be achieved in different ways, depending on the available data describing past and present biodiversity and ecosystem conditions. Most approaches assume that past ecosystem immigration–extinction dynamics were at equilibrium before a given change in ecosystem conditions (Kuussaari et al. 2009). Historical data can thus be used to estimate relationships linking biodiversity and ecosystem abiotic and/or biotic conditions (e.g. habitat quality, quantity, connectivity, abundance of keystone species or invasive predator) at equilibrium. Then, comparing present-day biodiversity and ecosystem conditions with the expected relationship at equilibrium makes it possible to detect potential extinction debts (Figure 21.3). When no data about past biodiversity is available, a widespread method is to use species–area relationships to evaluate if current biodiversity patterns match better with past than current habitat quantity and connectivity, to reveal ongoing extinction debt (Helm et al. 2005). If no historical data are available at all for the landscape of interest, extinction debt can also be detected by comparing present-day biodiversity patterns and ecosystem conditions with those from another similar landscape having a more stable history (Vellend et al. 2006). However, these methods do not provide direct estimations. Explicitly quantifying extinction debts (i.e. the expected number of future extinctions) is possible if past and present data are available for both biodiversity and ecosystem conditions. The combination of past and present data can be used to parameterize mechanistic models to assess current and future ecosystem dynamics according to varying landscape scenarios (May et al. 2013; Blanchard et al. 2020). Such data can also be used to fit species distribution models in order to evaluate and predict lags in species extinction and colonization following environmental changes (Blois et al. 2013; Talluto et al. 2017).

Box 21.2. *Detecting and predicting extinction debts from historical data*

Linking past ecosystem conditions and current biodiversity can still fail to unambiguously detect extinction debts (see Box 21.2), and, while rarely available, only long-term monitoring of biodiversity makes it possible to explicitly track species extinction dynamics (Kuussaari et al. 2009). For instance, Jones et al. (2018) illustrate how decades of time-series data makes it possible to relate local species extinction dynamics to landscape management history. Alternative sources of historical data also help to reconstruct past trajectories of biodiversity patterns and dynamics over longer periods (Vellend et al. 2013). Natural history collections offer many opportunities to assess past composition of ecosystems (Johnson et al. 2011). For instance, most herbarium collections contain records spanning over the entire 20th century and beyond, which can be used to evaluate past geographical distributions and phenology of species, or to estimate the dates of species invasions or extinction (Sax and Gaines 2008; Johnson et al. 2011; Vellend et al. 2013). Over larger time-scales, palynological, paleoecological records can help to track past changes in biodiversity and land cover in relation to human influences (Vellend et al. 2013; Bürgi et al. 2017). Such data can serve both to estimate past biodiversity state and past biotic conditions (e.g. abundance of keystone species) of ecosystems.

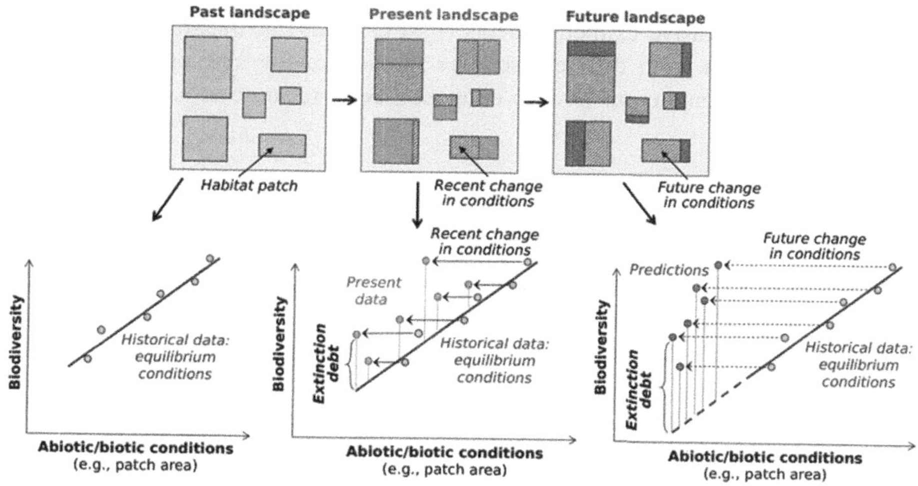

Figure 21.3. *General approach for evaluating extinction debt from landscape history. In the example here, a given landscape with six habitat patches is subject to recent changes in patch-level conditions (e.g. habitat loss). The relationship between biodiversity and ecosystem conditions at equilibrium is deduced from historical data (see Box 21.2). This relationship is then used to estimate extinction debt following changes in ecosystem conditions. The mismatch between this relationship at equilibrium and current biodiversity and ecosystem conditions (e.g. patch area) indicates an extinction debt. The amplitude of extinction debts associated with future fragmentation scenarios can be forecasted accordingly. For a color version of this figure, see www.iste.co.uk/decocq/ecology.zip*

Inertia in species extinction dynamics along with biotic dependency across species can entail co-extinction debts. Historical data provide valuable information about how changes in ecosystem conditions trigger delayed co-extinction. Numerous historical cases of co-extinction have been documented (Estes et al. 2011; Essl et al. 2015), including ancient trophic collapses (Roopnarine 2006). For instance, paleolimnology provides helpful quantitative tools to evaluate how past environmental changes have affected the structure of trophic networks in freshwater ecosystems (Jeppesen et al. 2001). These data can help to understand which combination of environmental change, characteristics of species (e.g. specific traits and trophic position) and ecosystem properties (food web structure, functional diversity), can entail co-extinction dynamics (Pillai et al. 2010; Brodie et al. 2014). Conclusions drawn from historically documented ecosystems where extinction debt has been partly or totally paid should be used, albeit with care, to make generalized predictions for similar and more recently affected systems. Notably, past ecosystem history can help to identify early warning signals of incipient critical transitions associated with pervasive extinction debts (Essl et al. 2015). Integrating more and more historical data from different sources should provide new opportunities to improve our understanding of how co-extinction debts relate to ecosystem legacies (Bürgi et al. 2017).

21.4. Implications for biodiversity conservation and management

While extinction debts are undoubtedly a challenge for biodiversity conservation, their delayed nature can let enough time to take measures preventing, or at least mitigating, future losses of biodiversity and ecosystem services (Figure 21.2, Vellend et al. 2006). This can be achieved through rapid and relevant restoration of habitat characteristics and ecosystem functions, before the debt is paid (Hylander and Ehrlén 2013; Cronk 2016). In addition, habitat areas with long-term environmental stability have high conservation values, as they are likely to act as refugia for remnant populations, and so as biodiversity reservoirs for future ecosystem restoration (Box 21.3). Increasing habitat quantity and connectivity should help to prevent the extinction of remnant populations. This can be done by identifying most spatially and environmentally suitable areas for habitat restoration (Newmark et al. 2017). However, the effects of restoration actions are likely to be delayed via an immigration credit, and this should be considered in conservation planning (Jackson and Sax 2010; Cronk 2016). When extinction debt arises from decreasing habitat quality, solutions should focus on maintaining or improving environmental conditions, and reducing stress factors. In the context of climate change, densifying vegetation canopy and reducing grazing pressure can help to maintain plant species populations (Vaughan and Gotelli 2021). Favoring species regional persistence by preserving and restoring thermally buffered areas (e.g. topographic concavities) might also mitigate climatic extinction debts (Lenoir et al. 2017).

Biological conservation is still mostly focused on preserving or improving habitat quality and spatial structure to reduce species extinction (Brodie et al. 2018). Detecting extinction debts and scheduling conservation action for all species interacting in an ecosystem is difficult, if not impossible, and only a few most endangered species are generally prioritized. Switching towards ecosystem-focused conservation policies with a more explicit integration of species function and interactions should help to mitigate delayed (co-)extinction dynamics (Figure 21.2). Identifying and conserving species providing key and unique ecological functions should have substantial positive effects in maintaining or restoring ecosystem-level biodiversity and services (Brodie et al. 2014, 2018, see also Box 21.3). Here again, historical ecology can serve to retrospectively evaluate how biodiversity legacies relate to past changes in ecosystem functioning (Ordonez and Svenning 2020), and which management practices have succeeded or failed in maintaining essential ecosystem functions (Figure 21.2).

European and North American forests have been subject to important human-driven impacts during the last centuries resulting from land-use changes (e.g. logging, expansion or abandonment of agricultural areas). Because they rely on slow demography of trees and associated soil dynamics, their long-term dynamics is intimately linked to historical anthropogenic influence (Foster et al. 2003; Hermy and Verheyen 2007). The abundance and quality of long-term historical data on these forests revealed that they are particularly heterogeneous in terms of age and historical legacies. As a result, they generally undergo long-term transient dynamics associated with both extinction debts and immigration credits (Vellend et al. 2006; Bergès and Dupouey 2020). The study of old forests (i.e. forests with long-term stable conditions) versus recent forests provides valuable insights for understanding the links between ecosystem history and delayed ecological dynamics, which should guide adequate conservation measures (Hermy and Verheyen 2007). Old forest patches have a high conservation value because they contain an important pool of forest specialist plants that slowly colonize recently reforested areas, entailing an immigration credit in recent forests (Hermy and Verheyen 2007; Bergès and Dupouey 2020). Moreover, large old trees often act as ecosystem engineers that provide resources, microclimatic conditions and physical habitat for many other plant and animal species (Jones et al. 2018). Preserving forest areas containing high density of old trees should thus sustain trophic networks and ecosystem functions.

Box 21.3. *The case of old forests*

21.5. Conclusion

The influence of extinction debts is pervasive under rapidly changing environmental conditions and landscape structures. However, conservation baselines are generally determined from short-term socio-ecological contexts and issues that are mostly blind to longer-term and transient ecological dynamics (Crumley et al. 2017). It is thus crucial to build conservation and land management policies that better integrate the history and legacies of ecosystems (Bürgi et al. 2017; Crumley et al. 2017). Achieving this goal will require both the future development of multidisciplinary research frameworks and raising awareness of the inertia driving ecosystem dynamics, whereby historical ecology should play a key role.

21.6. References

Bergès, L. and Dupouey, J. (2020). Historical ecology and ancient forests: Progress, conservation issues and scientific prospects, with some examples from the French case. *J. Veg. Sci.*, 32(1).

Blanchard, G., Birnbaum, P., Munoz, F. (2020). Extinction–immigration dynamics lag behind environmental filtering in shaping the composition of tropical dry forests within a changing landscape. *Ecography*, 43(6), 869–881.

Blois, J.L., Zarnetske, P.L., Fitzpatrick, M.C., Finnegan, S. (2013). Climate change and the past, present, and future of biotic interactions. *Science*, 341(6145), 499–504.

Brodie, J.F., Aslan, C.E., Rogers, H.S., Redford, K.H., Maron, J.L., Bronstein, J.L., Groves, C.R. (2014). Secondary extinctions of biodiversity. *Trends Ecol. Evol.*, 29(12), 664–672.

Brodie, J.F., Redford, K.H., Doak, D.F. (2018). Ecological function analysis: Incorporating species roles into conservation. *Trends Ecol. Evol.*, 33(11), 840–850.

Bürgi, M., Östlund, L., Mladenoff, D.J. (2017). Legacy effects of human land use: Ecosystems as time-lagged systems, *Ecosystems*, 20(1), 94–103.

Cronk, Q. (2016). Plant extinctions take time. *Science*, 353(6298), 446–447.

Crumley, C.L., Lennartsson, T., Westin, A. (eds) (2017). *Issues and Concepts in Historical Ecology: The Past and Future of Landscapes and Regions*, 1st edition. Cambridge University Press, Cambridge.

Didham, R., Tylianakis, J., Gemmell, N., Rand, T., Ewers, R. (2007). Interactive effects of habitat modification and species invasion on native species decline. *Trends Ecol. Evol.*, 22(9), 489–496.

Essl, F., Dullinger, S., Rabitsch, W., Hulme, P.E., Pyšek, P., Wilson, J.R.U., Richardson, D.M. (2015). Delayed biodiversity change: No time to waste. *Trends Ecol. Evol.*, 30(7), 375–378.

Estes, J.A., Terborgh, J., Brashares, J.S., Power, M.E., Berger, J., Bond, W.J., Carpenter, S.R., Essington, T.E., Holt, R.D., Jackson, J.B.C. et al. (2011). Trophic downgrading of planet earth. *Science*, 333(6040), 301–306.

Figueiredo, L., Krauss, J., Steffan-Dewenter, I., Sarmento Cabral, J. (2019). Understanding extinction debts: Spatio–temporal scales, mechanisms and a roadmap for future research. *Ecography*, ecog.04740.

Foster, D., Swanson, F., Aber, J., Burke, I., Brokaw, N., Tilman, D., Knapp, A. (2003). The importance of land-use legacies to ecology and conservation. *BioScience*, 53(1), 77–88.

Gilbert, B. and Levine, J.M. (2013). Plant invasions and extinction debts. *Proc. Natl. Acad. Sci.*, 110(5), 1744–1749.

Helm, A., Hanski, I., Partel, M. (2005). Slow response of plant species richness to habitat loss and fragmentation. *Ecol. Lett.*, 9(1), 72–77.

Hermy, M. and Verheyen, K. (2007). Legacies of the past in the present-day forest biodiversity: A review of past land-use effects on forest plant species composition and diversity. *Ecol. Res.*, 22(3), 361–371.

Hylander, K. and Ehrlén, J. (2013). The mechanisms causing extinction debts. *Trends Ecol. Evol.*, 28(6), 341–346.

Jackson, S.T. and Sax, D.F. (2010). Balancing biodiversity in a changing environment: Extinction debt, immigration credit and species turnover. *Trends Ecol. Evol.*, 25(3), 153–160.

Jeppesen, E., Leavitt, P., De Meester, L., Jensen, J.P. (2001). Functional ecology and palaeolimnology: Using cladoceran remains to reconstruct anthropogenic impact. *Trends Ecol. Evol.*, 16(4), 191–198.

Johnson, K.G., Brooks, S.J., Fenberg, P.B., Glover, A.G., James, K.E., Lister, A.M., Michel, E., Spencer, M., Todd, J.A., Valsami-Jones, E. et al. (2011). Climate change and biosphere response: Unlocking the collections vault. *BioScience*, 61(2), 147–153.

Jones, G.M., Keane, J.J., Gutiérrez, R.J., Peery, M.Z. (2018). Declining old-forest species as a legacy of large trees lost. *Divers. Distrib.*, 24(3), 341–351.

Jung, M., Rowhani, P., Scharlemann, J.P.W. (2019) Impacts of past abrupt land change on local biodiversity globally, *Nat. Commun.*, 10(1), 5474.

Kuussaari, M., Bommarco, R., Heikkinen, R.K., Helm, A., Krauss, J., Lindborg, R., Öckinger, E., Pärtel, M., Pino, J., Rodà, F. et al. (2009). Extinction debt: A challenge for biodiversity conservation. *Trends Ecol. Evol.*, 24(10), 564–571.

Lenoir, J., Hattab, T., Pierre, G. (2017). Climatic microrefugia under anthropogenic climate change: Implications for species redistribution. *Ecography*, 40(2), 253–266.

MacArthur, R. and Wilson, E. (1967). *The Theory of Island Biogeography*. Princeton University Press, Princeton.

Malanson, G.P. (2008) Extinction debt: Origins, developments, and applications of a biogeographical trope. *Prog. Phys. Geogr.*, 32(3), 277–291.

May, F., Giladi, I., Ristow, M., Ziv, Y., Jeltsch, F. (2013a). Metacommunity, mainland-island system or island communities? Assessing the regional dynamics of plant communities in a fragmented landscape. *Ecography*, 36(7), 842–853.

Mouquet, N., Matthiessen, B., Miller, T., Gonzalez, A. (2011). Extinction debt in source-sink metacommunities. *PLOS ONE*, 6(3), e17567.

Newmark, W.D. (1995) Extinction of mammal populations in western north american national parks. *Conserv. Biol.*, 9(3), 512–526.

Newmark, W.D., Jenkins, C.N., Pimm, S.L., McNeally, P.B., Halley, J.M. (2017). Targeted habitat restoration can reduce extinction rates in fragmented forests. *Proc. Natl. Acad. Sci.*, 114(36), 9635–9640.

Ordonez, A. and Svenning, J.-C. (2020). The potential role of species and functional composition in generating historical constraints on ecosystem processes. *Glob. Ecol. Biogeog.*, 29(2), 207–219.

Pillai, P., Loreau, M., Gonzalez, A. (2010). A patch-dynamic framework for food web metacommunities. *Theor. Ecol.*, 3(4), 223–237.

Roopnarine, P.D. (2006). Extinction cascades and catastrophe in ancient food webs. *Paleobiology*, 32(1), 1–19.

Rumpf, S.B., Hülber, K., Wessely, J., Willner, W., Moser, D., Gattringer, A., Klonner, G., Zimmermann, N.E., Dullinger, S. (2019). Extinction debts and colonization credits of non-forest plants in the European Alps. *Nat. Commun.*, 10, 4293.

Sax, D.F. and Gaines, S.D. (2008). Species invasions and extinction: The future of native biodiversity on islands. *Proc. Natl. Acad. Sci.*, 105(Suppl. 1), 11490–11497.

Talluto, M.V., Boulangeat, I., Vissault, S., Thuiller, W., Gravel, D. (2017). Extinction debt and colonization credit delay range shifts of eastern North American trees. *Nat. Ecol. Evol.*, 1(7), 0182.

Terborgh, J., Nuñez-Iturri, G., Pitman, N.C.A., Valverde, F.H.C., Alvarez, P., Swamy, V., Pringle, E.G., Paine, C.E.T. (2008). Tree recruitment in an empty forest. *Ecology*, 89(6), 1757–1768.

Tilman, D., May, R.M., Lehman, C.L., Nowak, M.A. (1994). Habitat destruction and the extinction debt. *Nature*, 371(6492), 65–66.

Valiente-Banuet, A., Aizen, M.A., Alcántara, J.M., Arroyo, J., Cocucci, A., Galetti, M., García, M.B., García, D., Gómez, J.M., Jordano, P. et al. (2015). Beyond species loss: The extinction of ecological interactions in a changing world. *Funct. Ecol.*, 29(3), 99–307.

Vaughan, I.P. and Gotelli, N.J. (2021). Using climatic credits to pay the climatic debt. *Trends Ecol. Evol.*, 36(2), 104–112.

Vellend, M., Verheyen, K., Jacquemyn, H., Kolb, A., Calster, H.V., Peterken, G., Hermy, M. (2006). Extinction debt of forest plants persists for more than a century following habitat fragmentation. *Ecology*, 87(3), 542–548.

Vellend, M., Brown, C.D., Kharouba, H.M., McCune, J.L., Myers-Smith, I.H. (2013). Historical ecology: Using unconventional data sources to test for effects of global environmental change. *Am. J. Bot.*, 100(7), 1294–1305.

22

Historical Ecology for the Past and the Future: Organizing at Local and Regional Scales

Carole L. CRUMLEY[1,2]

[1] *University of North Carolina, Chapel Hill, USA*

[2] *Swedish University of Agricultural Sciences, Uppsala, Sweden*

22.1. Introduction

Multiple unfolding crises menace not just humanity but all life on Earth. National, international and global institutions are moving too slowly to make a difference in the outcome for future generations. In the past two decades, the inability to come to agreements – and keep them – has stalled multinational progress towards goals that affect the entire planet (Clemençon 2016, Maizland 2021). This impasse has expanded scholarly and public interest in initiatives at local and regional scales; these can bring new energy and ideas to sluggish transnational scales and influence broader constituencies (e.g. Raja and Simpson 2021).

The action list is long: increase the pool of expertise by engaging all relevant knowledge communities, collect rapidly disappearing data, analyze with both familiar and new methods and apply actionable science to policy and practice (Palmer 2012; Beier et al. 2017). This enormously complex and urgent activity requires integrated research frameworks with the flexibility to accommodate the global diversity of places, peoples and processes, and the means to examine future options. Rather than waiting for funds and instructions, easily adaptable on-the-ground research designs with broad applicability can better guide practical efforts.

Historical Ecology,
coordinated by Guillaume DECOCQ. © ISTE Ltd 2022.

The *Integrated History and Future of People on Earth* (IHOPE) is an early example of how such initiatives by researchers have powered collaborations with diverse stakeholders, including residents and citizen scientists, local and regional administrative bodies, and commercial entities. The IHOPE example is particularly useful because it has benefitted from members' individual and stakeholder collaboration. In this fashion, the motor of development of durable practice is in the hands of those most knowledgeable of conditions, limitations and opportunities for action.

22.2. Founding IHOPE

Umbrella international scientific organizations were not particularly common in the 1980s. During its three decades of existence, the International Geosphere–Biosphere Programme (IGBP, 1986–2015), to better understand the Earth system, built research networks, facilitated synthesis and enhanced capacity around the world. The IGBP led a global effort to study how planet Earth functions and how it has – and will – change. The IGBP's framework linked land, sea and sky projects where they meet: land–atmosphere, land–ocean and atmosphere–ocean. The IGBP's clear global goals attracted 78 national committees, solidly representing every continent, and supported in-country collaborations through conferences and workshops; these encouraged both established researchers and new scientists to frame their research to address global challenges (IGBP 2015).

The Global Change Open Science Conference held in Amsterdam in 2001 was jointly organized by the IGBP, the International Human Dimensions Programme on Global Environmental Change (IHDP), the Global Programme on Biodiversity (DIVERSITAS) and the World Climate Research Programme (WCRP). The conference produced the "Amsterdam Declaration", which affirmed that the extent and impact of anthropogenic forces are equal to those of nature and called for an "ethical framework for global stewardship and strategies for Earth system management" (Moore III et al. 2001). This marked the global first steps towards including human activity in the analysis of Earth system research, a goal shared across the spectrum of IGBP and IHDP global programs.

At an IGBP conference in Banff, Canada in 2003, ecological economist Robert Costanza, a member of the IGBP task force Analysis, Integration and Modelling of the Earth System (AIMES), conceived the idea of constructing a timeline for human interaction with the Earth system. Will Steffen – an Earth system scientist and then executive director of the IGBP – agreed, viewing the initiative as consonant with the IGBP's aim to provide a holistic understanding of planetary dynamics.

Trained as an archaeologist and environmental anthropologist, and a member of the IGBP community (PAGES, AIMES), I quickly learned about the idea, soon named the *Integrated History and Future of People on Earth* (IHOPE). My archaeology colleagues and I were thrilled at the prospect of collaborating on such a timeline, as it aligned with our efforts to build a framework that would enable better integration of the Earth system sciences with the social sciences and humanities.

Keenly aware that we could not simply slap together different datasets and that modeling the past has its own challenges, our group began to expand in numbers and expertise. A Dahlem Conference was held in 2005, resulting in an edited book with contributions from a broad range of scholars and titled *Sustainability or Collapse? An Integrated History and Future of People on Earth* (Costanza et al. 2007a). This book, along with articles, meetings and workshops (including workshops at NCEAS), described the modeling and scenario building exercise and the framework for data integration (Costanza et al. 2007b, 2012; Heckbert et al. 2016). The IHOPE project office was first established in the United States (2007–2009; Hibbard et al. 2010), before relocating to Sweden (2009–present). After the dissolution of the IGBP in 2015, IHOPE became a global research project of a new global program, Future Earth (https://futureearth.org/).

22.3. Integrating the social sciences and humanities

The discussion between those who model and those who dig – and some who do both – was fascinating and at times contentious. The modelers wanted to begin their work right away, but the archaeologists and historians wanted to first establish a protocol for data collection and analysis that would smoothly meld the diverse perspectives and sources of evidence. In the end, there was a collective effort to identify drivers and monitoring tools that could render the archaeological record legible.

Trained to work with colleagues from many different disciplines and with many types of information, the archaeologists led the integrating exercise. As the *Annales* historian Marc Bloch (1993 [first published in 1949]) puts it:

> Very few scholars can boast that they are equally well equipped to read critically a medieval charter, to correctly interpret the etymology of placenames, to date unerringly the remains of prehistoric habitations, and to analyze the plants characteristic of a pasture, a field, or a moor. Without all these, however, how could one pretend to write the history of land-use? (author's translation).

Current archaeological practice is not only designed to assist collaboration among differently trained researchers but, in some national traditions, to also engage stakeholders in the area under study. This is particularly the case in North America where, despite brutal colonial histories, many indigenous peoples remain and are intimately tied to the land. North American archaeologists trained in anthropology departments have, over the last century, slowly moved archaeological practice towards ethical positions as regards indigenous rights (NAGPRA 1990) and greater awareness of the contemporary context of the past.

The research framework for IHOPE thus provides a kind of intellectual "contact zone" (Pratt 1991), in which diverse communities can exchange information about the past and the present of a physical region and discuss various courses of action for its future. The disciplines that have slowly moved together into this zone are anthropology, archaeology, history, geography and ecology. As in any contact zone, such exchanges require a strong commitment to dialogue, extensive translation and significant efforts at coordination, but they are feasible (Jones et al. 2018; Southerland-Wilson et al. 2019; Hicks 2021). In addition to a complex systems approach and attention to scales of time and space, IHOPE includes hybrid fields of study, such as bioarchaeology and critical cultural studies.

22.4. Historical ecology

Historical ecology is a cluster of concepts that offers a holistic, practical perspective to the study of environmental change. It may be applied to spatial and temporal frames at any resolution, finding particularly rich data sources at what is loosely termed the "landscape" scale, where human activity and biophysical systems interact and archaeological historical, and ethnographic records are plentiful. The term assumes a definition of ecology that includes humans as a component of all ecosystems, and a definition of history that encompasses the history of the Earth system as well as the social and physical past of our species (Balée 2006; Crumley 2007, 2013, 2015, 2016; Crumley et al. 2018). Historical ecology is not simply the study of the past, but the application of knowledge gained from its study to the human future.

The term seems to have almost been "in the air" beginning later in the 20th century (Deevey et al. 1979[1]; Bilsky 1980; Brooks 1985). Working in France in the mid-1970s, my colleagues and I struggled to give a name to the history of dynamic environments, finally adopting historical ecology to characterize our study of long-term regional change (Crumley and Marquardt 1987). Our subsequent

1. Deevey was using the term historical ecology as early as 1964.

collaboration with ecologists, archaeologists and anthropologists at a 1990 School for Advanced Research (SAR) workshop resulted in the publication of *Historical Ecology: Cultural Knowledge and Changing Landscapes* (Crumley 1994).

Independent of our work, the anthropologist and ethnobotanist William Balée and colleagues, working in South America, enriched traditional cultural ecology to include a strong historical approach to botany. Balée gathered anthropologists, historians and geographers in a 1994 workshop; the papers were published as *Advances in Historical Ecology* (Balée 1998; see also Balée 2006; Balée and Erickson 2006). Working together since the mid-1990s, Balée and Crumley have furthered the approach with the creation of publishing venues for historical ecology, at first with Columbia University Press and now with Routledge.

Historical ecology is predicated on the assumption that it is possible to construct evidence-validated narratives of landscape transformation that has resulted from the continual interaction of diverse human activities with changing environmental conditions. Historical ecology perforce draws on a broad spectrum of concepts, methods, theories and evidence, taken from the biological and physical sciences, ecology, the social sciences and the humanities. Derived from the standard procedures of various disciplines, independent datasets provide critical "cross-checks" upon one another, revealing patterns of association with researchers and often yielding further intriguing questions.

Due to the Earth's enormous environmental and cultural diversity, no single research strategy is possible. Instead, the flexibility of the historical ecology "toolbox" encourages the matching of available kinds of information with local and regional conditions (Meyer and Crumley 2011). For example, both tropical and boreal research settings are rich in evidence of past conditions, but the tools for their treatment are necessarily different, in response to pragmatic and strategic issues such as preservation, funding, stakeholders and access.

Many of the participants in the 1990 SAR workshop became members of the IHOPE Scientific Steering Committee, so it is not surprising that the IHOPE collective embraced the historical ecology framework. Attracted by the possibilities offered by this framework, individuals with textual, artistic, linguistic and other expertise from the environmental humanities joined the group.

In the last decade, IHOPE has increased attention to several areas. Climate history (Ljungqvist et al. 2021), modeling and scenario building (Scarborough and Isendahl 2020; Scarborough 2021), landscape research (Crumley et al. 2017), and the powerful approach of complex systems thinking (Sinclair et al. 2018) are only a few of IHOPE contributions that advance understanding of landscapes, regions and cultural change. Most recently, IHOPE has completed a book that examines ancient urban regions and applies our findings to the future (Murphy and Crumley 2022).

22.5. Conclusion

The role of global programs has changed over the last 20 years, from facilitating research in the Earth system sciences to facilitating their integration with long-term human cognition, activity and behavior. In 2018, the International Council for Science (ICSU) and the International Social Science Council (ISSC) merged, now the International Science Council (ISC). Among the newest of international sustainability programs is the UNESCO BRIDGES global coalition that aims to integrate humanities, social science and other knowledges into research, education and action for global sustainability "at local and territorial scales" (UNESCO 2019). This on-the-ground inclusion of human activity as a driver of global change underscores the importance of seeing the Earth as a complex system.

Strengthened with a more comprehensive assessment of the issues, we can better understand and respond to crisis in multiple domains. However, contrasting global goals and national endeavors and diminished national and international funding will curtail the geographical reach of global programs. The unraveling of the Paris Accord and other setbacks have accelerated the tendency to shift scales/arenas of action from broad global goals to regional and local endeavors, such as those of IHOPE.

The opportunity to meet and work at the global scale with researchers from disciplines beyond one's own was exciting and led to IHOPE's founding. IHOPE's contribution is to facilitate work at the local and regional levels that can build towards global goals. IHOPE is well placed to offer a smooth path among scales, from historical global climates and environments to those to come.

There is no single solution to today's challenges, because global natural, cultural and historical diversity is too great. Instead, we must work from the bottom up: from local and regional projects to broader scales, with flexible frameworks and a variety of research tools. A good example is the necessary rethinking of one-size-fits-all agricultural practices that use the same management tools across different landscapes, biomes, soils, climates/microclimates and heritage. The historical ecology toolbox, a complex systems approach, and the shifting needs and conditions of the future call for on-the-ground flexibility.

This requires a collaboration between scholars and diverse stakeholders. How to expand the research you do to the next village, county or countryside? Now is the time to join with like-minded colleagues in your region. Find them at conferences, contact those whose work inspires you, plan a project within your administrative precinct(s), explore alliances with interested regional partners, find funding together by increasing your networks. In addition to regional collaborations, the new concepts and ideas that you have can be shared and expanded; this is how IHOPE came to be.

You may say that would be nice but there is no funding. But for many of the last 18 years, IHOPE has operated *pro bono*, with no employees or website manager, powered instead by a group of colleagues. With sporadic funding since the dissolution of the IGBP, we have maintained the essentials, keeping a website running since 2013, becoming affiliated with similar endeavors and using colleagues' extensive networks to build a vibrant community. IHOPE researchers, using our framework, have been quite successful in finding funding for their own research. We feature their work and that of others who offer their projects on the website portal; the website itself is widely used as a teaching tool. Thanks to the Internet, it is possible to build and maintain a coalition without extensive funding.

For IHOPE veterans, the shared activity of building a framework began with the dialogue between archaeologists and natural scientists; for the archaeologists, the historical ecology workshop at SAR gave us confidence that the IHOPE framework could work in very different places. Your ideas and your energy can carry us into a liveable future.

22.6. References

Balée, W. (ed.) (1998). *Advances in Historical Ecology*. Columbia University Press, New York.

Balée, W. (2006). The research program of historical ecology. *Annu. Rev. Anthropol.*, 35(5), 15–24.

Balée, W.L. and Erickson, C.L. (2006). *Time and Complexity in Historical Ecology: Studies in the Neotropical Lowlands*. Columbia University Press, New York.

Beier, P., Hansen, L.J., Helbrecht, L., Behar, D. (2017). A how-to guide for coproduction of actionable science. *Conserv. Lett.*, 10(3), 288–296.

Bilsky, L.J. (ed.) (1980). *Historical Ecology: Essays on Environment and Social Change*. Kennikat Press, Port Washington, NY.

Bloch, M. (1953). *The Historian's Craft*. Vintage, New York.

Bloch, M. (1993). *Apologie pour l'histoire ou métier d'historien*. Armand Colin, Paris.

Brooks, D.R. (1985). Historical ecology: A new approach to studying the evolution of ecological associations. *Ann. Missouri Bot. Garden*, 72(4), 660–680.

Clémençon, R. (2016). The two sides of the Paris climate agreement: Dismal failure or historic breakthrough? *J. Env. Dev.*, 25(1), 3–24.

Costanza, R.L., Graumlich, L.J., Steffen, W. (eds) (2007a). *Sustainability or Collapse? An Integrated History and Future of People on Earth*. Dahlem Workshop Reports # 96. MIT Press, Cambridge, MA.

Costanza, R.L., Graumlich, L., Steffen, W., Crumley, C.L., Dearing, J., Hibbard, K., Leemans, R., Redman, C., Schimel, D. (2007b). Sustainability or collapse: What can we learn from integrating the history of humans and the rest of nature? *Ambio*, 36, 522–527.

Costanza, R.L., van der Leeuw, S., Hibbard, K., Aulenbach, S., Brewer, S., Burek, M., Cornell, S., Crumley, C.L., Dearing, J., Folke, C. et al. (2012). Developing an integrated history and future of people on Earth (IHOPE). *Curr. Opin. Env. Sustain.*, 4, 106–114.

Crumley, C.L. (ed.) (1994). *Historical Ecology: Cultural Knowledge and Changing Landscapes*. School of American Research, Santa Fe.

Crumley, C.L. (2007). Historical ecology: Integrated thinking at multiple temporal and spatial scales. In *The World System and the Earth System: Global Socio-environmental Change and Sustainability Since the Neolithic*, Hornborg, A. and Crumley, C.L. (eds). Left Coast Press, Walnut Creek, CA.

Crumley, C.L. (2013). Historical ecology in archaeology. In *Encyclopedia of Global Archaeology*, Smith, C. (ed.). Springer, New York

Crumley, C.L. (2015). New paths into the Anthropocene. In *Handbook of Historical Ecology and Applied Archaeology*, Isendahl, C. and Stump, D. (eds). Oxford University Press, Oxford.

Crumley, C.L. (2016). Historical ecology. In *The Wiley Blackwell International Encyclopedia of Anthropology*, Callan, H. (ed.). Wiley Blackwell, New York.

Crumley, C.L. and Marquardt, W.H. (eds) (1987). *Regional Dynamics: Burgundian Landscapes in Historical Perspective*. Academic Press, San Diego.

Crumley, C.L., Kolen, J.C.A., de Kleijn, M., van Manen, N. (2017). Studying long-term changes in cultural landscapes: Outlines of a research framework and protocol. *Landsc. Res.*, 42(8), 880–890.

Crumley, C.L., Lennartsson, T., Westin, A. (eds) (2018). *Issues and Concepts in Historical Ecology: The Past and Future of Landscapes and Regions*. Cambridge University Press, Cambridge.

Deevey, E.S., Rice, D.S., Rice, P.M., Vaughan, H.H., Brenner, M., Flannery, M.S. (1979). Mayan urbanism: Impact on a tropical karst environment. *Science*, 206, 298–306.

Heckbert, S., Costanza R.L., Parrott L. (2016). Achieving sustainable societies: Lessons from modelling the ancient Maya. *Solutions*, 5(5), 55–64.

Hibbard, K.A., Costanza R., Crumley C., van der Leeuw S., Aulenbach, S., Dearing J., Morais J., Steffen, W., Yasuda, Y. (2010). Developing an integrated history and future of people on earth (IHOPE): Research plan. Global Change. IGBP Report No. 59. IGBP Secretariat, Stockholm.

Hicks, M. (2021). How archeology harms and helps indigenous resistance: Lessons from Wet'suwet'en [Online]. Available at: https: //www.gc.cuny.edu/ [Accessed 20 March 2021].

IGBP (2015). National Committee List [Online]. Available at: http://www.igbp.net/people/nationalcommittees/nationalcommitteelist.4.950c2fa1495db7081e183f5.html [Accessed 26 March 2021].

Jones, E.A., Westin, A., Madry, S., Murray, S., Moen, J., Tickner, A. (2018). How to operationalise collaborative research. In *Issues and Concepts in Historical Ecology: The Past and Future of Landscapes and Regions*, Crumley, C.L., Lennartsson, T., Westin, A. (eds). Cambridge University Press, Cambridge.

Ljungqvist, F.C., Seim, A., Huhtamaa, H. (2021). Climate and society in European history. *Wires Clim. Change*, 12, e691.

Maizland, L. (2021). Global climate agreements: Successes and failures [Online]. Available at: https://www.cfr.org/backgrounder/paris-global-climate-change-agreements [Accessed 27 March 2021].

Meyer, W.J. and Crumley, C.L. (2011). Historical ecology: Using what works to cross the divide. In *Atlantic Europe in the First Millennium BC: Crossing the Divide*, Moore, T. and Armada, L. (eds). Oxford University Press, Oxford.

Moore III, B., Underdal, A., Lemke, P., Loreau, M. (2001). Amsterdam declaration on Earth system science [Online]. Available at: http://www.igbp.net/about/history/2001amsterdamdeclarationonearthsystemscience.4.1b8ae20512db692f2a680001312.html [Accessed 7 April 2021].

Murphy, J.T. and Crumley, C.L. (eds) (2022). *If the Past Teaches, What Does the Future Learn? Ancient Urban Regions and the Durable Future.* Research in Urbanism Series, Delft School of Architecture, TU Delft OPEN Publishing, Delft.

NAGPRA (1990). The Native American Graves Protection and Repatriation Act (NAGPRA). Pub. L. 101-601, 25 U.S.C. 3001 et seq., 104 Stat. 3048, is a United States Federal Law enacted on November 16, 1990.

Palmer, M.A. (2012). Socioenvironmental sustainability and actionable science. *BioScience*, 62(1), 5–6.

Pratt, M. (1991). Arts of the contact zone. *Profession*, 1991, 33–40.

Raja, R. and Simpson, I.A. (2021). Scaling up and zooming in: Global history and high-definition archaeology perspectives on the longue durée of urban–environmental relations in Gerasa (Jerash, Jordan). *J. Global Hist.*, 16(1), 1–20.

Scarborough, V.L. (2021). What ancient landscapes contribute to climate change. *Econ. Anthropol.*, 8, 161–167.

Scarborough, V.L. and Isendahl, C. (2020). Distributed urban network systems in the tropical archaeological record: Towards a model for urban sustainability in the era of climate change. *Anthropocene Rev.*, 7(3), 1–23.

Sinclair, P., Moen, J., Crumley, C.L. (2018). Historical ecology and the longue durée. In *Issues and Concepts in Historical Ecology: The Past and Future of Landscapes and Regions*, Crumley, C.L., Lennartsson, T., Westin, A. (eds). Cambridge University Press, Cambridge.

Southerland-Wilson, D., Spice, A., Armstrong, C.G. (2019). Compliance archaeology fails indigenous people in British Columbia: An example from Unist'ot'en territory. *The Midden*, 49(1), 26–29.

UNESCO (2019). Toward the establishment of Bridges: Action to promote sustainability science [Online]. Available at: https://en.unesco.org/news/toward-establishment-bridges-action-promote-sustainability-science [Accessed 8 April 2021].

List of Authors

Vincent BALLAND
Bourgogne-Franche-Comté University
Dijon
France

Christophe BALTZINGER
INRAE
UR EFNO
Nogent-sur-Vernisson
France

Antoine BECKER-SCARPITTA
University of Helsinki
Finland
and
Institute of Botany of the Czech Academy of Science
Průhonice
Czech Republic

Grégoire BLANCHARD
University of Montpellier
France

Boris BRASSEUR
Jules Verne University of Picardie
Amiens
France

Jérôme BURIDANT
Jules Verne University of Picardie
Amiens
France

Clémence CHAUDRON
University of Lorraine
Nancy
France

Roger COLY
University of Tours
France

Marco CONEDERA
Swiss Federal Research Institute WSL
Cadenazzo
Switzerland

Carole L. CRUMLEY
University of North Carolina
Chapel Hill
USA
and
Swedish University of Agricultural Sciences
Uppsala
Sweden

Pieter DE FRENNE
Ghent University
Belgium

Emiel DE LOMBAERDE
Ghent University
Belgium

Karen DE PAUW
Ghent University
Belgium

Pallieter DE SMEDT
Ghent University
Belgium

Guillaume DECOCQ
Jules Verne University of Picardie
Amiens
France

Francesca DI PIETRO
University of Tours
France

Alexa DUFRAISSE
Muséum National d'Histoire Naturelle
Paris
France

Ove ERIKSSON
Stockholm University
Sweden

Damien ERTLEN
University of Strasbourg
France

Elisabet FERNÁNDEZ-CABRERA
Universidad de Las Palmas de Gran Canaria
Canary Islands
Spain

Fabrizio FERRETTI
CREA Research Centre for Forestry and Wood
Arezzo
Italy

Catherine FRUCHART
Bourgogne-Franche-Comté University
Besançon
France

Emilie GALLET-MORON
Jules Verne University of Picardie
Amiens
France

Leví GARCÍA-ROMERO
Universidad de Las Palmas de Gran Canaria
Canary Islands
Spain

Frank S. GILLIAM
University of West Florida
Pensacola
USA

Olivier GIRARDCLOS
Bourgogne-Franche-Comté University
Besançon
France

Karen GREIG
University of Otago
Dunedin
New Zealand

Luis HERNÁNDEZ-CALVENTO
Universidad de Las Palmas de
Gran Canaria
Canary Islands
Spain

Antonio Ignacio
HERNÁNDEZ-CORDERO
Universidad de Las Palmas de
Gran Canaria
Canary Islands
Spain

Pierre-Alexis HERRAULT
University of Strasbourg
France

Michael Timm HOFFMAN
University of Cape Town
South Africa

Hélène HOREN
Jules Verne University of Picardie
Amiens
France

Isabelle JOUFFROY-BAPICOT
Bourgogne-Franche-Comté University
Besançon
France

Keith J. KIRBY
University of Oxford
UK

Patrik KREBS
Swiss Federal Research Institute WSL
Cadenazzo
Switzerland

Jonathan LENOIR
Jules Verne University of Picardie
Amiens
France

Samuel LETURCQ
University of Tours
France

Karl-Johan LINDHOLM
Uppsala University
Sweden

Alberto MALTONI
University of Florence
Italy

Damien MARAGE
Bourgogne-Franche-Comté University
Besançon
France

Anders MÅRELL
INRAE
UR EFNO
Nogent-sur-Vernisson
France

Giorgio MARESI
Fondazione Edmund Mach
San Michele all'Adige
Italy

Néstor MARRERO-RODRIGUEZ
Universidad de Las Palmas de
Gran Canaria
Canary Islands
Spain

Justin J. MAXWELL
Sunrise Archaeology
Mangonui
New Zealand

Camille MEEUSSEN
Ghent University
Belgium

François MUNOZ
Grenoble Alpes University
Grenoble
France

Claudia OLIVEIRA
University of Lorraine
Metz
France

Bastien PARISY
University of Helsinki
Finland

Carolina PEÑA-ALONSO
Universidad de Las Palmas de
Gran Canaria
Canary Islands
Spain

Emma PÉREZ-CHACÓN ESPINO
Universidad de Las Palmas de
Gran Canaria
Canary Islands
Spain

Giovanna PEZZI
University of Bologna
Italy

Nicolas J. RAWLENCE
University of Otago
Dunedin
New Zealand

Vincent ROBIN
University of Lorraine
Metz
France

Rick F. ROHDE
University of Cape Town
South Africa

Tomas ROSLIN
University of Helsinki
Finland
and
Swedish University of Agricultural
Sciences
Uppsala
Sweden

Ian D. ROTHERHAM
Sheffield Hallam University
UK

Aarón Moisés SANTANA-CORDERO
Universidad de Las Palmas de
Gran Canaria
Canary Islands
Spain

David SHEEREN
University of Toulouse
France

Lara D. SHEPHERD
Museum of New Zealand Te Papa
Tongarewa
Wellington
New Zealand

Thomas VANNESTE
Ghent University
Belgium

Kris VERHEYEN
Ghent University
Belgium

Alexander J.F. VERRY
University of Otago
Dunedin
New Zealand
and
University of Toulouse
France

Donald M. WALLER
University of Wisconsin
Madison
USA

Richard WALTER
University of Otago
Dunedin
New Zealand

Florian ZELLWEGER
Swiss Federal Institute for Forest,
Snow and Landscape Research
Birmensdorf
Switzerland

Index

E, F

G, H

I, L

V, W

Printed and bound by CPI Group (UK) Ltd, Croydon, CR0 4YY
13/09/2026
14960997-0007